Healing herbs of India

Healing Herbs of India

by :

ANUBHA UPADHYAYA
PREETI SAGAR NAYAK
V.K. AGRAWAL
A.B. TIWARI

Department of Crop and Herbal Physiology
Jawaharlal Nehru Krishi Vishwa Vidyalaya, Jabalpur - 482 004
Niche Area of Excellence in - Conservation, Cultivation, Processing and Quality Evaluation of Medicinal & Aromatic Plants

2009

SATISH SERIAL PUBLISHING HOUSE

403, Express Tower, Commercial Complex, Azadpur, Delhi-110033 (India)
Phone : 011-27672852, Fax : 91-11-27672046
E-mail :info@satishserial.com, hkjain1975@yahoo.com
Website : www.satishserial.com

Published by :

SATISH SERIAL PUBLISHING HOUSE

403, Express Tower, Commercial Complex, Azadpur, Delhi-110033 (INDIA)
Phone : 011-27672852 Fax : 91-11-27672046
E-mail : info@satishserial.com, hkjain1975@yahoo.com

ISBN 81-89304-72-0

Typeset at: Typographiya, Delhi # 9811482328
Printed at: Salasar Imaging Systems, Delhi-35

PREFACE

From time unknown Indian medicinal plants have been used to cure and prevent ailments. The pharmacological values of many of these plants have been confirmed by scientific findings. Medicinal and Aromatic plants are rich sources of secondary metabolites like alkaloids, glycosides, steroids, flavoinoids, saponins etc which are potential sources of modern herbal drugs. It is possible that the efficiency of herbs depend on the total cumulative effect of the plant contents rather than on the one or few chemical fractions separated from the herbs.

The authentic information regarding use of medicinal plants, active ingredients, pharmacological properties and dosage etc are not available at one place. An efforts thus has been made by us to compile the systematic and proven information in the form of a book.

This book in particular caters to the need of all those who are trying to decide whether to use herbs as a form of treatment or an adjunctive treatment or as health promoter or simply as a prophylactic measure for disease prevention and immunity building.

The World Health Organization (WHO) recognizes that medicinal and aromatic plants have an important role to play in our scientifically and technologically advanced world. The Department of Crop and Herbal Physiology, College of Agriculture, JNKVV, Jabalpur has been recognized as centre of excellence on medicinal and aromatic plants. The basic purpose of publishing this book is to provide a well documented reference on the identification, distribution, chemical composition and usage of medicinal and aromatic plants.

Chapter **2**

Arthritis

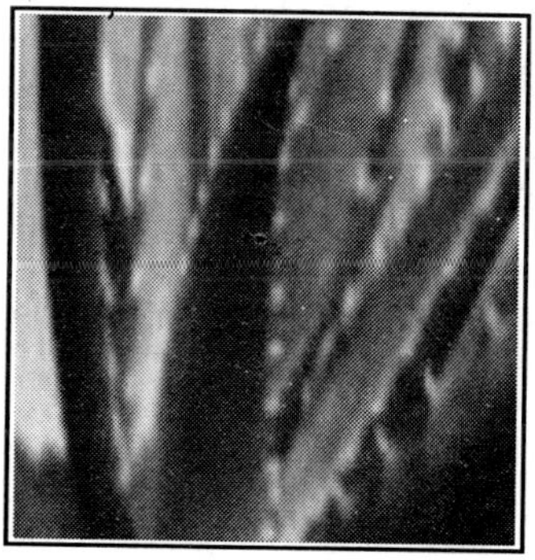

Botanical Name : *Aloe barbadensis*

Family : Liliaceae

Common Name : Ghi kanvar

Parts used : Leaves

Active ingredient : Barbaloin, Aloesin, Aloesone, Aloe-emodin

Pharmacological properties : Tonic, Purgative

Dosages : Extract two teaspoons juice from aloe leaf. Mix with 6-8 drops of lemon juice and ½ teaspoon of Jira powder. Apply on the joints once a day.

Botanical Name : *Allium sativum*

Family : Alliaceae

Common Name : Lahsun

Parts used : Bulb

Active ingredient : Volatile oil, Alliin, Allinase, Vitamins-Thiamene, Riboflavin and Niacin

Pharmacological properties : Anti inflammatory, Anodyne, Diuretic, Anti spasmodic

Dosages : Boil 6-7 cloves of Lahsun in 150ml milk for 2-3 minutes. Drink at night for 5-10 days.

Botanical Name: *Acorus calamus*

Family : Araceae

Common Name : Buch

Parts used : Rhizomes

Active ingredient : α and β-Asarone, Acoric acid, Choline

Pharmacological properties : Analgesic, Sedative, Stimulant, Tonic

Dosages : Rub a piece of rhizomes on the sand stone with a few drops of warm water to make paste. Apply on the painful joints once in a day.

Botanical Name : *Calotropis gigantea*

Family : Asclepiadaceae

Common Name : Madar

Parts used : Leaves, Flowers

Active ingredient : Calotropins, Calatropageni, Sapogenins

Pharmacological properties : Astringent, Analgesic, Tonic

Dosages :

1. Mix juice of 5-6 leaves equal quantity of Til oil. Apply on affected part once in a day.
2. Powder dried flowers. Take once pinch of the powder with little jaggery twice a day for four to six days.

Botanical Name : *Cyperus rotundus*

Family : Cyperacae

Common Name : Motha

Parts used : Rhizome

Active ingredient : Sesquiterpenoids, Sugeonol, Cyperol, Copaene, Mustakone

Pharmacological properties : Anti inflammatory, Astringent, Analgesic, Cooling

Dosages : Powder ½ rhizome and 2gms dry ginger. Mix powder in a glass of warm water. Consume twice a day when pain is severe.

Botanical Name : *Commiphora wightii*

Family : Burseraceale

Common Name : Guggal

Parts used : Resin

Active ingredient : Guggulipid

Pharmacological properties : Anti inflammatory, Astringent, Analgesic, Anti arthritic

Dosages : Grind 60gms of guggal, 20gms of Amla, 20gms of each of Bahera and Harra in 100 ml of water. Consume ½ teaspoon daily.

Botanical Name: *Datura stramonium*

Family : Solanaceae

Common Name : Datura

Parts used : Leaves

Active ingredient : Hyoscine, Hyoscyamine

Pharmacological properties : Diuretic, Anti rheumatic, Anti spasmodic, Narcotic

Dosages : Take 8-10 leaves and put in any hot oil till they turn black. Strain oil and use for massaging. Massage for 15 days twice a week.

Botanical Name : *Emblica officinalis*

Family : Euphorbiaceae

Common Name : Amla

Parts used : Fruits

Active ingredient : Vitamin C, Tannins, Gallic acid, Ellagic acid, Phyllemblic acid, and Emblicol

Pharmacological properties : Thermogenic, Anodyne, Rejuvenating, Tonic

Dosages : Grind 20gms fruit of amla, 20g each of Bahera and Harra and 60gms of guggal in 100 ml of water. Consume ½ teaspoon per day.

Botanical Name: *Pongamia pinnata*

Family : Fabaceae

Common Name : Karanj

Parts used : Leaves

Active ingredient : Karanjin, Kanjone

Pharmacological properties : Analgesic

Dosages : Boil handful of leaves in just enough water to prepare a paste. Apply on affected parts once in a day.

Botanical Name : *Piper longum*

Family : Piperaceae

Common Name : Pipali

Parts used : Seeds

Active ingredient : Piperine

Pharmacological properties : Anti pyretic, Anodyne, Purgative, Anti inflammatory

Dosages : Grind 10gms of pipali and 10gms of salt to make fine powder. Consume 5gms of powder with warm water in the morning for 20 days.

Botanical Name : *Papaver somniferum*

Family : Papaveraceae

Common Name : Aphim

Parts used : Latex

Active ingredient : Narcotine, Morphine, Papaverine

Pharmacological properties : Diuretic, Astringent, Alkaloids-Morphine and Codeine, Potent analgesic

Dosages : Collect a very small amount (0.25 to 0.50gms) of dried latex. Consume 2-3 times a day.

Botanical Name : *Ricinus communis*

Family : Euphorbiaceae

Common Name : Arandi

Parts used : Leaves

Active ingredient : Amino acids

Pharmacological properties : Tonic, Anti inflammatory, Anodyne

Dosages : Put any cooking oil on the leaves, heat them and apply on affected joints. Use once a day for one week.

Botanical Name : *Sida cordifolia*

Family : Malvaceae

Common Name : Bariarai

Parts used : Root

Active ingredient : γ-Ephedrine, Ephedrine, Vasicinol, Vasicinone

Pharmacological properties : Diuretic, Astringent, Tonic

Dosages : Prepare decoction with 2gms of root powder with 100ml of water. Add to this a pinch of dry adrak powder. Consume the whole quantity once a day.

Botanical Name : *Solanum nigrum*

Family : Solanaceae

Common Name : Makoi

Parts used : Leaves, Stem

Active ingredient : Spirostanol, Glycosides, Furostanol

Pharmacological properties : Tonic, Luxative, Anti inflammatory, Analgesic

Dosages : Extract juice from leaves and stems. Slightly warm the juice. Apply on painful joints once a day.

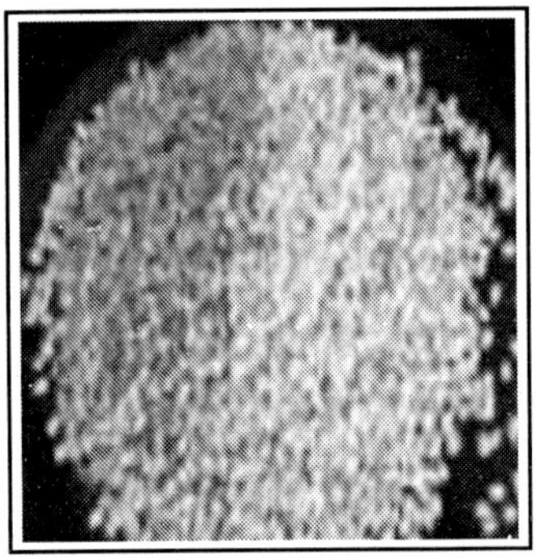

Botanical Name : *Sesamum indicum*

Family : Pedaliaceae

Common Name : Til

Parts used : Seeds

Active ingredient : Sterols, Sesamin, Sesamolin, Sesamol

Pharmacological properties : Thermog-enic, Emmolient, Anti arrhritic, Tonic

Dosages : Crush 20gms of seeds. Extract juice from 2" of fresh adrak. Remove solid parts and store the extract in a bottle. Apply on affected parts once in a day.

Botanical Name: *Tribulus terrestris*

Family : Zygophyllaceae

Common Name : Gokhuru

Parts used : Whole plant

Active ingredient : Steroidal Sapogenins, Saponins, Flavonoids, Carboline alkanoids

Pharmacological properties : Anti inflammatory, Tonic

Dosages : Crush whole plant of gokhrur to extract juice. Strain juice and equal quantity of water. Boil till water reduces to ¼ in quantity. To this decoction add ¼ quantity of sesamum oil and boil on low flame till only the oil remains. Store the oil. Apply oil on the joints twice a day.

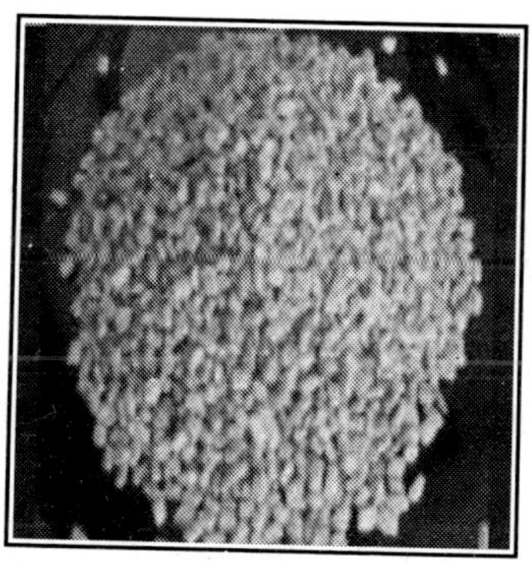

Botanical Name : *Trigonella foenum graecum*

Family : Fabaceae

Common Name : Methi

Parts used : Seeds

Active ingredient : Sapogenins, Saponins, Fenugreekine

Pharmacological properties : Thermog-enic, Anti arthritic, Tonic

Dosages :

1. Mix methi (10gms) with rice (100gms) to make Khichdi. Eat two times a day.
2. Mix 20gms of powdered seeds in water. Drink once a day.

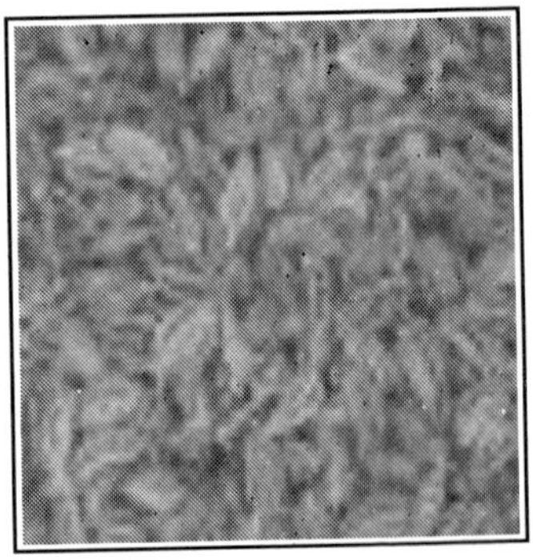

Botanical Name : *Trachyspermum ammi*

Family : Apiaceae

Common Name : Ajowan

Parts used : Seeds

Active ingredient : Coumarins, β–sitosterol

Pharmacological properties : Thermogenic, Digestive, Anti rheumatic

Dosages : Grind 20gms seeds and 20gms of dry ginger to make fine powder. Consume 10gms powder in one glass of warm water in the morning for 40 days.

Botanical Name: *Vitex negundo*

Family : Verbenaceae

Common Name : Nirgundi

Parts used : Leaves

Active ingredient : Nishindine, flavonoids

Pharmacological properties : Astringent, Sedative, Anti arthritic

Dosages :

1. Prepare paste with 5-10 leaves. Apply paste on affected parts once in a day.
2. Wrap 5gms leaves and 30gms common salt in a muslin cloth and heat it for application. Apply on painful joints 2-3 times per day.

— xxx —

Chapter 3

Asthma

Botanical Name : *Terminalia arjuna*

Family : Combretaceae

Common Name : Arjun

Parts used : Bark

Active ingredient : Arjungenin, Baicalein, Arjunolone

Pharmacological properties : Demulcent, Expectorant, Alexeteric

Dosages :

1. Make powder of bark put honey in it and take early in the morning once a day.
2. Soak bark in one glass water overnight. Drink infusion in the morning on empty stomach till the problem persists.

Botanical Name : *Piper betle*

Family : Piperaceae

Common Name : Paan

Parts used : Leaves

Active ingredient : β and γ-sitosterols, Hentriacontane, Pentatriacontane, Stearic and Chavicol-Carvacrol, Eugenol, Chavicol, Allyl catechol, Cadinine.

Pharmacological properties : Analgestic, Stimulant, Expectorant

Dosages : Dip 1-2 betel leaves in warm mustard oil enough to cover the leaves. Massage oil on chest before going to bed.

— xxx —

Chapter **4**

Bleeding from Nose

Botanical Name : *Rose centifolia*

Family : Rosaceae

Common Name : Gulab

Parts used : Flowers

Active ingredient : Gallic acid

Pharmacological properties : Cooling

Dosages : Take 20-25 red rose petals and mix with one table spoon of sugar. Keep it in an air tight bottle for few days till it turns into a smooth paste called Gulukand. Take 15 gms gulukand with 50 gms curd 2-3 times in a day.

Botanical Name: *Ocimum gratissium*

Family : Lamiaceae

Common Name : Rama Tulsi

Parts used : Leaves

Active ingredient : Ursolic acid, β-sitosterol, Cholesterol, Stigmasterol, Fatty acids

Pharmacological properties : Alexeteric

Dosages : Add 4-5 tulsi leaves, half teaspoon tea leaves, two cardamom and half teaspoon sugar in 15 ml of water and boil it. After boiling add milk in it and boil again. Drink 3-4 times in a day.

— xxx —

Chapter 5

Constipation

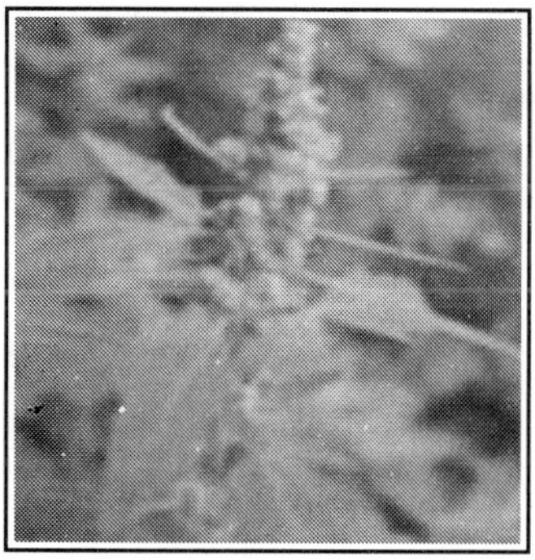

Botanical Name : *Chenopodium album*

Family : Chenopodiaceae

Common Name : Bathua

Parts used : Leaves

Active ingredient : Kaempferol-7-rhamoniside and Ambroside

Pharmacological properties : Nutritious, Luxative

Dosages : Boil 15-20 Bathua leaves in one glass of water. Strain it and mix one tsp of lemon juice. Drink one glass once a day for 2-3 days.

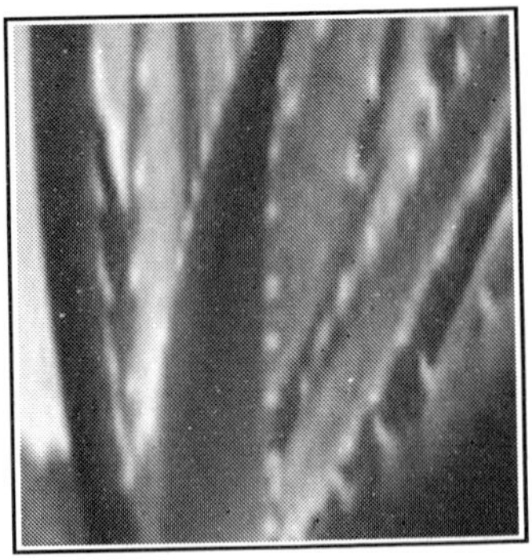

Botanical Name : *Aloe barbedensis*

Family : Liliaceae

Common Name : Ghikanvar

Parts used : Leaves

Active ingredient : Barbaloin, Aloesin, Aloesone, Aloe-emodin

Pharmacological properties : Anthelmimtic, Aperient, Carminative, Digestive, Stomachic

Dosages : Grind pulp of 6-8 Ghikuamr leaves to make paste. Add a pinch of salt and turmeric and mix well. Take ½ teaspoon of this paste twice a day for 2-3 days.

Botanical Name : *Bacopa monnieri*

Family : Scrophulariaceae

Common Name : Brahmi

Parts used : Leaves

Active ingredient : Herpestine

Pharmacological properties : Luxative, Car minative

Dosages : Boil 10-15 thin leaves in one glass of water. Add a pinch of salt to the solution. Drink one glass twice a day for 2-3 days.

Botanical Name : *Ficus benghalansis*

Family : Moraceae

Common Name : Bargad

Parts used : Fruit

Active ingredient : β-sitosterol, Rutin

Pharmacological properties : Astringent, Luxative, Tonic

Dosages : Soak one fruit of Bargad in a cup of milk for overnight. Drink once a day for 2-3 days before going to sleep.

Botanical Name :*Asparagus racemosus*

Family : Asparagaceae

Common Name : Satavar

Parts used : Root

Active ingredient : Shatavarins, Saponins, Asparinins A and B, Asparoside A and B

Pharmacological properties : Astringent, Anti-spasmodic, Luxative, Refrigerant, Tonic

Dosages : Boil root of wild satavar in 100ml of water. Take this liquid with little sugar once in a day for one week.

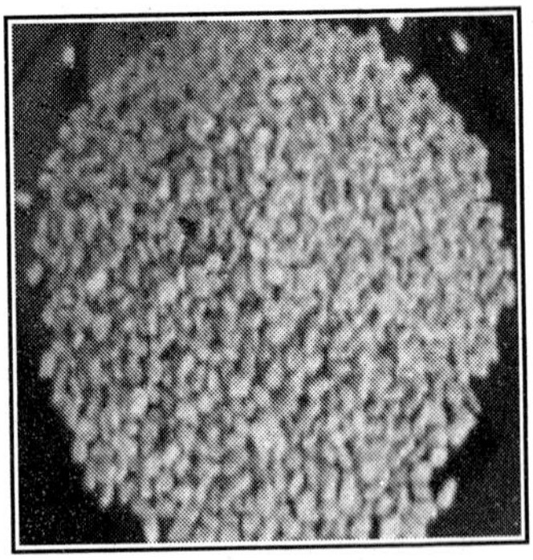

Botanical Name : *Trigonella foenum graecum*

Family : Facaceae

Common Name : Methi

Parts used : Seeds

Active ingredient : Sapogenins, Saponin, Fenugreekine

Pharmacological properties : Demulcent, Stomachic,Tonic, Supportive

Dosages : Make dry vegetable of Palak and green Methi. Take one katori once a day for 2-3 days.

Botanical Name : *Foreniculum vulgare*

Family : Apiaceae

Common Name : Sounf

Parts used : Seeds

Active ingredient : Flavonoids, Umbelliferone

Pharmacological properties : Digestive, Stimulant, Luxative

Dosages : Take 1 tsp seeds with one glass of water once in a day for 1-2 days.

Botanical Name : *Swertia chirayita*

Family : Gentianaceae

Common Name : Chirayita

Parts used : Whole plant

Active ingredient : Carvacrol, v-terpinine & p-cymene, α & β pinene, Camphene, Myrcene, α-terpinene

Pharmacological properties : Stomachic, Anthelmintic, Luxative

Dosages : Boil the whole plant in sufficient water till the quantity reduces to half. Take 1-2 tsp of decoction 2-3 times a day till the problem persists.

Botanical Name : *Tinospora cardifolia*

Family : Menispermaceae

Common Name : Giloy

Parts used : Leaves

Active ingredient : Cardifolioside A & B, β-Sitosterol, Tinosporin

Pharmacological properties : Digestive, Tonic, Carminative

Dosages : Dip 2-3 leaves in two tsp of ghee. Take once a day for 2-3 days.

Botanical Name :*Azadirachta indica*

Family : Meliaceae

Common Name : Neem

Parts used : Leaves

Active ingredient : Nimbin, Nimbolides, Nimocinolides, Isonimocinolides, Quercetin, β-sitosterol

Pharmacological properties : Anodyne, Anti-bacterial, Tonic

Dosages : Grind 1-2 tsp leaves to make paste and mix in one cup of water. Drink 1-2 tsp 4-5 times a day for 1-2 days

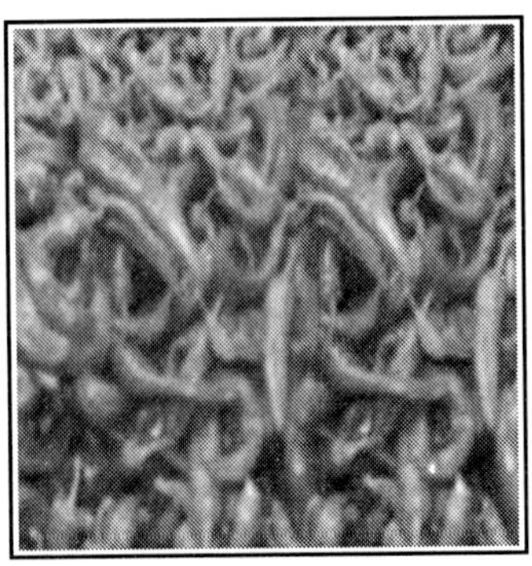

Botanical Name : *Capsicum annum*

Family : Solanaceae

Common Name : Harvi mirch

Parts used : Fruit

Active ingredient : Capsaicin, Homocapsaicin, Vanillylamide

Pharmacological properties : Digestive, Carminative, Stimulant, Stomachic

Dosages : Take 2-3 green chillies with meals for two days.

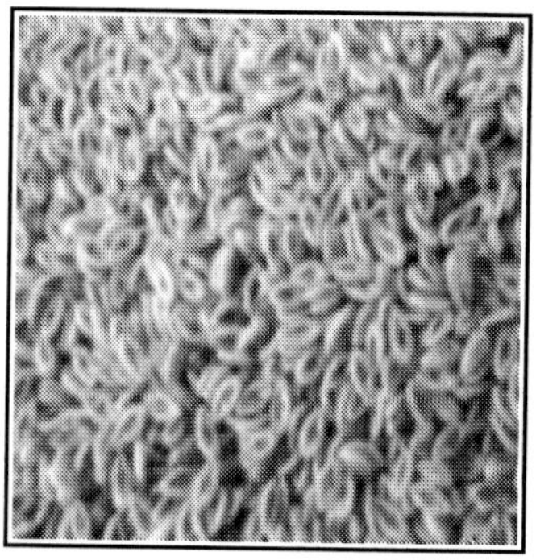

Botanical Name : *Plantago ovata*

Family : Plantaginaceae

Common Name : Isabgol

Parts used : Seeds

Active ingredient : Valine, Alanine, Glycine, Cystine, Leucine, Tyrosine

Pharmacological properties : Emollient, Demulcent, Luxative, Carminative

Dosages : Take one tsp husk of Isabgol and soak in water overnight. Take 2-3 times in a day for two days.

— xxx —

Chapter 6

Cough and Cold

Botanical Name : *Ocimum gratissium*

Family : Lamiaceae

Common Name : Rama Tulsi

Parts used : Leaves

Active ingredient : Ursolic acid, β-sitosterol, Cholesterol, Stigmasterol, Fatty acids

Pharmacological properties : Demulcent, Expectorant, Alexeteric

Dosages : Crush 6-8 tulsi leaves and grind with 10gms ginger. Extract juice from both. Mix well and add half teaspoon of honey. Consume 3-4 drops twice a day for a week

Botanical Name : *Zingiber officinale*

Family : Zingiberaceae

Common Name : Adrak

Parts used : Rhizomes

Active ingredient : Asparogine, Pipecolic acid, Free amino acids

Pharmacological properties : Expectorant, Emollient, Anodyne, Appetizer

Dosages : Boil 5gms ginger, mulathee and cardamom each in one glass of water till it remains half glass. Take one fourth tea spoon with honey in the morning and evening for a week.

Botanical Name : *Glycyrrhiza glabra*

Family : Fabaceae

Common Name : Mulathee

Parts used : Root

Active ingredient : Glycyrrhizin, Glycyrrhizic acid

Pharmacological properties : Expectorant, Demulcent, Refrigerant, Emetic, Alexeteric, Alterant, Tonic.

Dosages :

1. Boil 1 piece of Mulathee in two glasses of water till only half of water is left. Drink twice a day for two days.
2. Grind 100 g Mulathee to make powder. Take 1/4th teaspoon of the powder and add 1 teaspoon honey to it. Consume with warm water once a day for 10 days.

Botanical Name : *Piper nigrum*

Family : Piperaceae

Common Name : Kali mirch

Parts used : Fruit

Active ingredient : Sabinene, Myrcene, Limonene and β-Pinene

Pharmacological properties : Alexeteric, Stimulant, Digestive

Dosages : To a glass of warm sweet milk add 2 pinch of black pepper. Drink once for 2 days.

Botanical Name : *Cannabis sativa*

Family : Cennabinaceae

Common Name : Bhang

Parts used : Leaves

Active ingredient : Flavonoid glycosides

Pharmacological properties : Anti-inflammatory, Analgesic, Thermogenic, Intoxicating

Dosages : Burn 20-25 dry leaves of bhang to convert them into ash. Store ash in airtight container. Pinch of powder twice a day with honey till the medicine finishes

Chapter 7

Diabetes

Botanical Name : *Azadirachta indica*

Family : Meliaceae

Parts used : Neem

Parts used : Leaves

Active ingredient : Nimbin, Nimbolides, Nimocinolides, Isonimocinolides, Quercetin, β-sitosterol

Pharmacological properties : Hypogly-cemic, Antibiotic, Tonic

Dosages : Boil 10 leaves in 1 cup of water for five to ten minutes and strain the liquid. Drink one cup daily.

Botanical Name : *Asparagus racemosus*

Family : Asparagaceae

Common Name : Satavar

Parts used : Roots

Active ingredient : Shatavarins, Saponins, Asparinins A and B, Asparoside A and B

Pharmacological properties : Bitter, Luxative, Stimulant, Tonic

Dosages : Powder 10gms dried roots of Satavar and 10gms of methi seeds. Take 20 gms daily for one month.

Botanical Name : *Alstonia scholaris*

Family : Apocynaceae

Common Name : Chatwan

Parts used : Stem

Active ingredient : Lupeol acetate, Amyrin acetate

Pharmacological properties : Bitter, Tonic, Anthelmintic, Cooling, Astringent

Dosages : Rub stem on sand stone with few drops of water to get one-teaspoon paste. Consume one teaspoon paste once a day.

Botanical Name : *Ficus benghalensis*

Family : Moraceae

Common Name : Bargad

Parts used : Bark

Active ingredient : Leucoanthocyanin, Leucoanthocyanidin

Pharmacological properties : Cooling, Tonic, Hypoglycemic

Dosages : Grind dried bark on sand stone to get powder. Take ½ teaspoon powder once daily for 3 - 4 days.

Botanical Name : *Gymnema sylvestre*

Family : Asclepiadaceae

Common Name : Gudmar

Parts used : Leaves

Active ingredient : Gymnemic acid, Hentriacomtanol

Pharmacological properties : Luxative, Hypoglycemic, Astringent, Carminative, Saponins and Tritertines

Dosages :

1. Grind handful of leaves and boil in ½ litres water to make ¼ litres of decoction. Take 1 cup per day for 40 days.
2. Prepare dried leaves powder. Take ½ teaspoon once in a day.

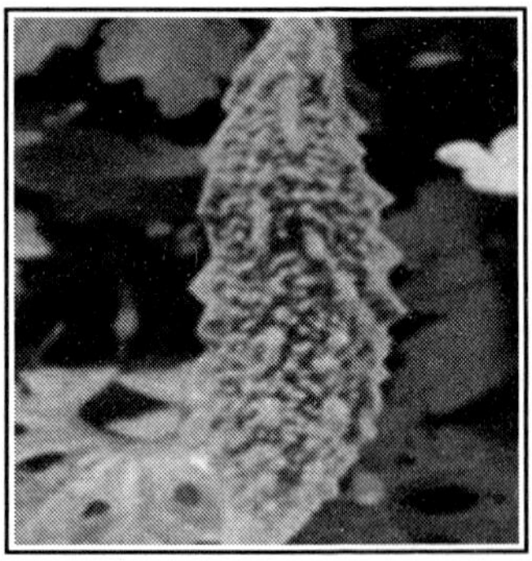

Botanical Name : *Mormordica charantia*

Family : Cucurbitaceae

Common Name : Karela

Parts used : Fruit

Active ingredient : Cucurbitacin glycosides

Pharmacological properties : Hypogl-ycemic, Anti diabetic, Regenerative, Tonic, Anti oxidant

Dosages : Extract juice from the Karela fruit by adding a little water. Take 10-20 ml juice in the morning in empty stomach.

Botanical Name : *Mangifera indica*

Family : Anacardiaceae

Common Name : Aam

Parts used : Leaves

Active ingredient : Mangiferin

Pharmacological properties : Regenerative, Bitter, Anthelmintic, Astringent

Dosages : Dry the leaves in shade and powder it to preserve for use when necessary. Take ½ teaspoon twice a day with water in the morning and evening.

Botanical Name : *Emblica officinalis*

Family : Euphorbiaceae

Common Name : Amla

Parts used : Fruit

Active ingredient : Vitamin C, Tannins, Gallic acid, Ellagic acid, Phyllemblic acid, and Emblicol

Pharmacological properties : Cardiac, Luxative, Tonic

Dosages : Extract 10ml fruit juice by crushing the fruit. Mix with a pinch of turmeric. Take once in a day in empty stomach

Botanical Name : *Pterocarpus marsupium*

Family : Fabaceae

Common Name : Bijasal

Parts used : Bark

Active ingredient : Marsupol, Carpusin, Pterostibine

Pharmacological properties : Cooling, Bitter, Astringent, Acrid, Hypoglycemic

Dosages : Boil 50gms of bark of Pterocartus marsupium, 50gms root of Casearia esculenta, 50gms of creeper of Tinospora cardifolia (Giloe), 50g bark of Syzygium cumini (Jira), in one litre of water to make ¼ litre of decoction. Drink ½ cup decoction twice a day for one month.

Botanical Name : *Ricinus communis*

Family : Euphorbiaceae

Common Name : Arandi

Parts used : Stem

Active ingredient : Amino acids

Pharmacological properties : Emollient, Depurative, Luxative, Soothing

Dosages : Sequeeze out juice by crushing fresh stem. Drink 50 ml juice once a day for 5 - 7 months.

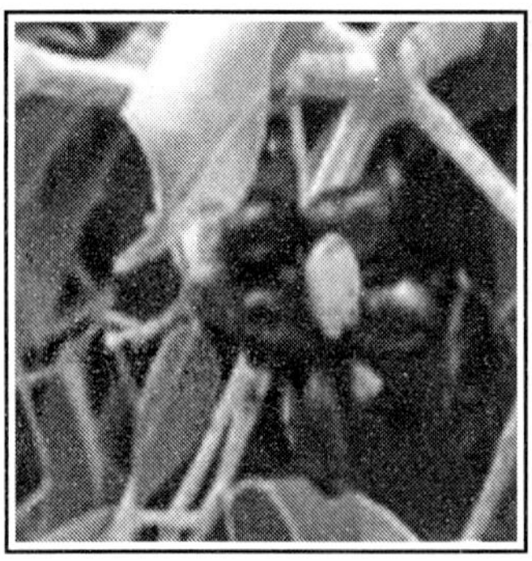

Botanical Name : *Syzygium cumini*

Family : Myrtaceae

Common Name : Jamun

Parts used : Seed

Active ingredient : β-sitosterol

Pharmacological properties : Glucoside-jamboline, Regenerative, Astringent, Carminative, Hypoglycemic

Dosages : Crush seeds and mix with little water. Strain to get the liquid. Take one teaspoon twice or thrice daily.

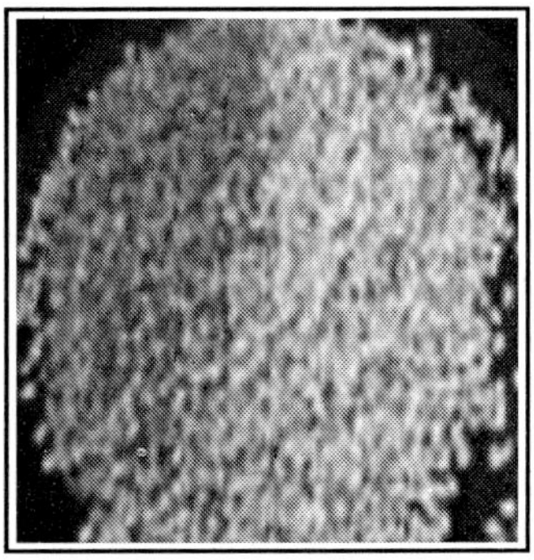

Botanical Name : *Sesamum indicum*

Family : Pedaliaceae

Common Name : Til

Parts used : Seeds

Active ingredient : Sterols, Sesamin, Sesamolin, Sesamol, Tocopherol

Pharmacological properties : Emollient, Demulcent, Luxative, Hypoglycemic

Dosages : Roast 3-5 gms of til seeds and grind with little milk. Take paste twice a day for one month.

Botanical Name : *Tinosporia cordifolia*

Family : Menispermaceae

Common Name : Giloy

Parts used : Stalk

Active ingredient : Tinosporin, Berberine

Pharmacological properties : Cooling, Digestion, Alexiteric, Anti diabetic, Alkaloids, Immunomodulator, Tonic

Dosages : Crush about 6" stalk (leaves and stem) to extract juice. Drink one cup juice daily.

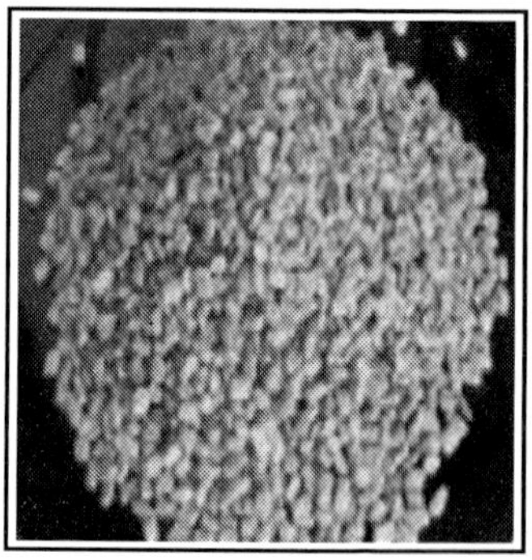

Botanical Name : *Trigonella foenum-graceum*

Family : Fabaceae

Common Name : Methi

Parts used : Seeds

Active ingredient : Sapogenins, Saponins, Steroidal Sapogenins, Fenugreekine

Pharmacological properties : Insulin sensitizer, High polyphenol, Alkaloid and Trigonelline, Saponins

Dosages : Take 25gms of seeds and boil in one cup of water. Drink one cup twice a day.

Botanical Name : *Terminalia chebula*

Family : Combretaceae

Common Name : Harra

Parts used : Fruits

Active ingredient : Chebulin, Chebulinic acid, Tannic acid

Pharmacological properties : Bitter, Anodye, Digestive, Tonic, Anti-diabetic

Dosages : Grind 10gms fruit of Terminalia bellirica (Bahera), Terminalia chebula and 10g fruit of Phyllanthus emblica (Amla) are to get powder. Take one teaspoon powder with water twice.

Botanical Name : *Vinca rosea*

Family : Apocynaceae

Common Name : Sadabahar

Parts used : Leaves

Active ingredient : Vinblastine, Vincristine, Vincaleukoblastine

Pharmacological properties : Tonic, Alkaloids, Tranquilising, Hypotensive, Anti diabetic

Dosages : Crush five leaves and two flowers to extract juice. Strain the juice. Take on alternate days i.e. three days / week, on alternate weeks for four weeks.

— xxx —

Botanical Name : *Coriandrum sativum*

Family : Apiaceae

Common Name : Dhania

Parts used : Seeds

Active ingredient : Linalool, Camphene, Geraniol, α & β-pinene, Myrcene

Pharmacological properties : Carminative, Refrigerant, Stomachic, Tonic

Dosages : Take 2-3 tsp coriander seeds and soak overnight. Grind and add to one cup of buter milk. Drink in the morning till symptom persists.

Botanical Name : *Tamarindus indica*

Family : Casalpiniaceae

Common Name : Imli

Parts used : Seeds

Active ingredient : Tartaric acid and Malic acid

Pharmacological properties : Astringent, Digestive, Carminative

Dosages : Grind tamarind (Imli) seeds to make powder. Add this paste to cup of water. Drink once a day.

Botanical Name : *Tinospora cardifolia*

Family : Menispermaceac

Common Name : Giloy

Parts used : Leaves

Active ingredient : Cardifolioside A & B, β- Sitosterol, Tinosporin

Pharmacological properties : Digestive, Tonic, Carminative

Dosages : Grind 10-15 leaves of Giloy to make paste. Add it to a glass of water. Drink once a day at night.

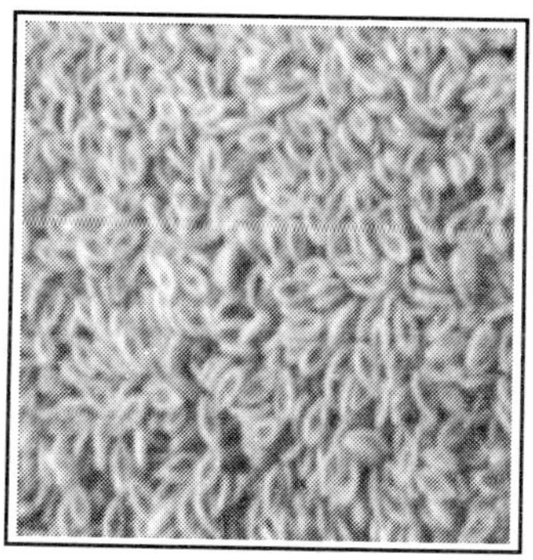

Botanical Name : *Plantago ovata*

Family : Plantaginaceae

Common Name : Isabgol

Parts used : Seeds

Active ingredient : Valine, Alanine, Glycine, Cystine, Leucine, Tyrosine

Pharmacological properties : Emollient, Demulcent, Cooling

Dosages : Mix two tsp seeds of Isabgol in half cup of curd. Take twice or thrice a day for three day.

Botanical Name : *Azadirachta indica*

Family : Meliaceae

Common Name : Neem

Parts used : Leaves

Active ingredient : Nimbin, Nimbolides, Nimocinolides, Isonimocinolides, Quercetin, β-sitosterol

Pharmacological properties : Antiseptic, Astringent, Cooling, Stimulant

Dosages : Grind 250gm each of leaves of Neem, Black plum and Bael. Make tiny tablets and dry. Take two tablets once in a day till problem subsides.

Botanical Name : *Phyllanthus emblica*

Family : Euphorbiaceae

Common Name : Amla

Parts used : Fruit

Active ingredient : Vitamin C, Tannins, Gallic acid, Ellagic acid, Phyllemblic acid, Emblicol

Pharmacological properties : Appetizer

Dosages : Grind 40-50 medium size leaves of Amla and mix in a glass of buttermilk, take once in a day.

— xxx —

Chapter 9

Dysentry

Botanical Name : *Mentha arvensis*

Family : Lamiaceae

Common Name : Pudina

Parts used : Leaves

Active ingredient : Linalool, Camphene, Geraniol, α & β - pinene, Myrcene

Pharmacological properties : Anthelmintic, Antiseptic, Carminative, Antispasmodic, Stimulant

Dosages : Grind leaves and strain to extract juice. Mix sugar to improve the taste. Take two teaspoon twice daily with a glass of water for two days.

Botanical Name : *Coriandrum sativum*

Family : Apiaceae

Common Name : Dhania

Parts used : Seeds

Active ingredient : Linalool, Camphene, Geraniol

Pharmacological properties : Stomachic, Stimulant, Anthelmintic, Carminative, Antispasmodic

Dosages : Grind fresh leaves to extract juice. Mix 1-2 teaspoon fresh juice in one cup of buttermilk. Drink this thrice a day for three days.

Botanical Name : *Saraca indica*

Family : Caesalpiniaceae

Common Name : Asoka

Parts used : Flowers

Active ingredient : β-sistosterol, Palimitic, Stearic, Linoleic acid

Pharmacological properties : Refrigerant, Astringent, Anthelmintic, Demulcent

Dosages : Grind 2-3 flowers of Asoka tree to take out juice and mix juice with a glass of water. Drink once a day.

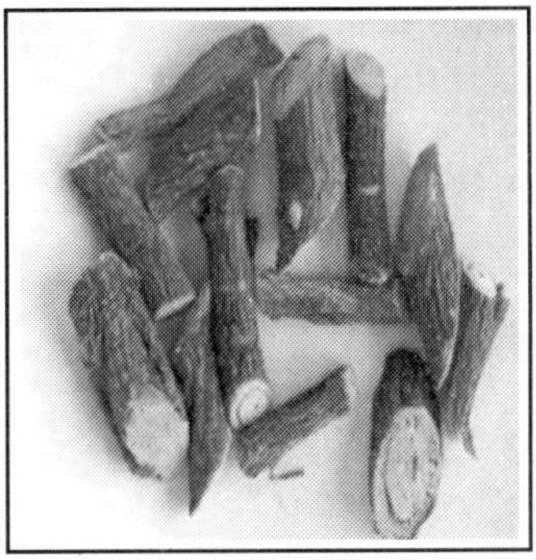

Botanical Name : *Glycyrrhiza glabra*

Family : Fabaceae

Common Name : Mulethi

Parts used : Twig

Active ingredient : Glycyrrhizin, Glycyrrhizic acid

Pharmacological properties : Demulcent, Antibiotic, Anti-bacterial, Anodyne

Dosages : Make paste of one piece of dry twig by rubbing on sandstone. Mix it with one teaspoon of honey. Take ½ teaspoon of this paste twice a day.

Botanical Name : *Butea monosperma*

Family : Fabaceae

Common Name : Dhak

Parts used : Stem bark

Active ingredient : Palasonin, β-sitosterol, α–amyrin, Palastrin, Butrin

Pharmacological properties : Anthelmintic, Astringent

Dosages : Grind stem bark of Dhak, Banraj (Bauthina racemosa) and gum of Drumstick (Moringa olefera) to make powder. Take one tsp of powder in the morning and evening daily for 2-3 days.

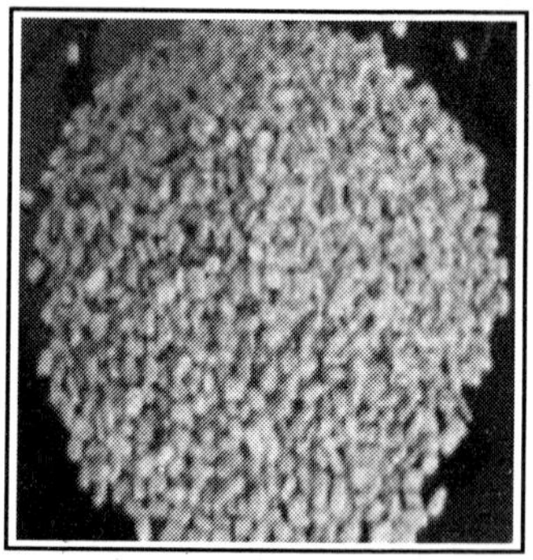

Botanical Name : *Trigonella foenum graecum*

Family : Fabaceae

Common Name : Methi

Parts used : Seeds

Active ingredient : Sapogenins, Saponin, Fenugreekine

Pharmacological properties : Anthelmintic, Astringent, Appetizer, Carminative

Dosages : Soak 8-10 Seeds of Methi in 2 teaspoon of ghee overnight. Take once a day for two days.

Botanical Name : *Lawsonia inermis*

Family : Lythraceae

Common Name : Henna

Parts used : Leaves

Active ingredient : Luteolin - sitosterol, Glucosides, Lawsone

Pharmacological properties : Constipating, Anthelmintic, Anti-bacterial

Dosages : Grind 4 teaspoon of Henna leaves to make powder. Mix with one tsp of ghee and make small balls Take one ball twice a day till problem subsides.

— XXX —

Chapter 10

Ear Infection

Botanical Name : *Allium sativum*

Family : Alliaceae

Common Name : Lahsun

Parts used : Bulb

Active ingredient : Volatile oil, Alliin, Allinase, Vitamins-Thiamene, Riboflavin and Niacin

Pharmacological properties : Anodyne, Thermogenic, Anti-bacterial, Anti-fungal, Alexeteric

Dosages : Heat 5-6 cloves of lahsun in 6 tea spoon of mustard oil. Til oil when lahsun is fried then sieve oil. Cool and store in a bottle. Put two drops every day in the ear.

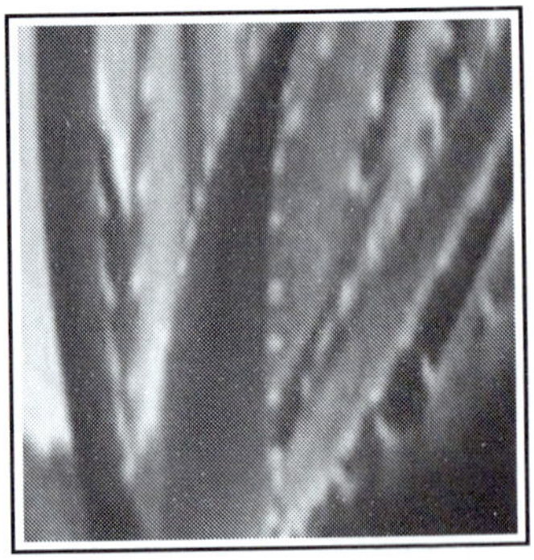

Botanical Name : *Aloe barbadensis*

Family : Liliaceae

Common Name : Ghikanvar

Parts used : Leaves

Active ingredient : Barbaloin, Aloesin, Aloesone, Aloe-emodin

Pharmacological properties : Alexeteric, Depurative

Dosages : Grind aloe leaves pulp and extract juice. Put two drops every day in the ear.

Botanical Name : *Acorus calamus*

Family : Araceae

Common Name : Buch

Parts used : Rhizomes

Active ingredient : α and β-Asarone

Pharmacological properties : Alexeteric, Anodyne, Anti-inflammatory

Dosages : Add powdered root of Buch and crushed garlic to coconut oil, heat for 2-3 minutes and cool. Put two drops of coconut oil in each ear once a day.

Botanical Name : *Ocimum basilicum*

Family : Lamiaceae

Common Name : Babui tulsi

Parts used : Leaves

Active ingredient : Eugenol, Geranial, Ocimene, Citronellol, Limonene, Tannins

Pharmacological properties : Anti-inflammatory, Anti-bacterial, Anodyne

Dosages : Grind few leaves and add to coconut oil. Heat the oil and then cool it. Put two drops of the oil in each ear once a day.

— xxx —

Chapter **11**

Eye Problems

Botanical Name : *Allium cepa*

Family : Alliaceae

Common Name : Piyaz

Parts used : Bulb

Active ingredient : Organic sulphur compounds including trans-S-L- cysteine sulphoxide. Quercetin, Vanillic acid

Pharmacological properties : Ophthalmic, Refrigerant

Dosages : Extract juice from one small Piyaz and mix equal amount of honey in it. Put one drop in each eye once a day.

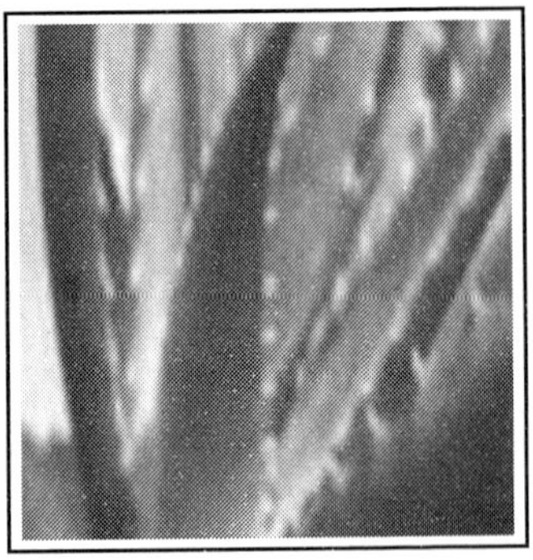

Botanical Name : *Aloe barba densis*

Family : Liliaceae

Common Name : Ghikanvar

Parts used : Leaves

Active ingredient : Barbaloin, Aloesin, Aloesone, Aloe-emodin

Pharmacological properties : Alexeteric, Ophthalmic, Cooling

Dosages : Peel off the skin and get the pulp of the leaf. Place the pulp on the eyes bandage, and keep overnight. Repeat it for 3 days.

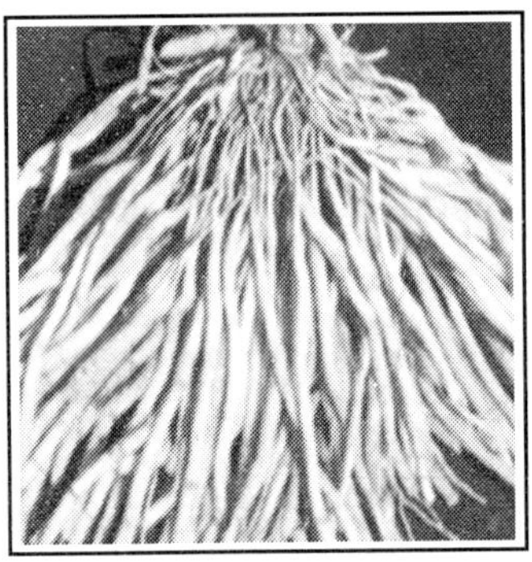

Botanical Name : *Asparagus racemosus*

Family : Asparagaceae

Common Name : Satavar

Parts used : Root

Active ingredient : Shatavarins, Saponins, Asparinins A and B, Asparoside A and B

Pharmacological properties : Ophthalmic, Cooling, Anodyne

Dosages : Boil 20 gm fresh root in one litre of water till one fourth quantity remains. Drink one cup twice a day.

Botanical Name : *Azadirahcta indica*

Family : Meliaceae

Common Name : Neem

Parts used : Leaves

Active ingredient : Quercetin, β-sitosterol, Nimbin, Nimbolides

Pharmacological properties : Antiseptic, Anti-inflammatory, Ophthalmic, Refrigerant

Dosages : Extract juice of 50-60 neem leaves. Mix equal amount of water. Wash eyes with this liquid twice a day for 5 days.

Botanical Name : *Butea mono sperma*

Family : Fabaceae

Common Name : Dhak

Parts used : Roots

Active ingredient : Palastrin, Butrin, Isobutrin Flavonoids, Glucosides

Pharmacological properties : Astringent, Anti-inflammatory, Anodyne, Ophthalmic

Dosages : Extract the juice from 2-3 roots and strain it. Put 1-2 drops thrice a day for 3 days.

Botanical Name : *Crocus sativum*

Family : Iridaceae

Common Name : Kesar

Parts used : Dry stigma

Active ingredient : Crocins, Crocetin, Picrotoxin

Pharmacological properties : Anti-inflammatory, Anodyne

Dosages : Take 2 gm Kesar and 2gm honey, mix them well. Apply on eyelids twice a day.

Botanical Name : *Croton tiglium*

Family : Euphorbiaceae

Common Name : Jamal gota

Parts used : Seeds

Active ingredient : Phorbol, Myristate acetate

Pharmacological properties : Anti-inflammatory, Ophthalmic

Dosages : Soak 10 gm seeds of Jamal gota in 20 gm lemon juice for 3 days and dry in sun. Grind to make fine powder. Apply as kajal in the eyes daily.

Botanical Name : *Adhatoda vasica*

Family : Acanthaceae

Common Name : Adusa

Parts used : Leaves

Active ingredient : Vasicine Vasicol, Adhatonine, Vasicinone, Vasicinol, Vasicinolone

Pharmacological properties : Astringent, Ophthalmic

Dosages : Take 10-15 leaves of Adusa nut tree. Grind to make paste. Bandage on eyes for 10-15 minutes.

Botanical Name : *Phyllanthus emblica*

Family : Euphorbiaceae

Common Name : Amla

Parts used : Fruit

Active ingredient : Vitamin C, Tannins, Gallic acid, Ellagic acid, Phyllemblic acid, Emblicol

Pharmacological properties : Astringent, Anti-inflammatory, Ophthalmic

Dosages : Take out juice from 50 gm fresh fruit by soaking the fruit and then grinding to extract juice. Put 2-3 drops in each eye thrice a day for 3 days.

Botanical Name : *Ficus banghalensis*

Family : Moraceae

Common Name : Bargad

Parts used : Leaves

Active ingredient : Leucoanthocyanin, Leucoanthocyanidin

Pharmacological properties : Anthelm-intic, Digestive

Dosages : Grind 10-15 fresh leave to extract juice and add a glass of water. Take once a day.

— xxx —

Chapter 13

General Health Problems

FEVER

Botanical Name : *Aegle marmelos*

Family : Rutaceae

Common Name : Bael

Parts used : Leaves & Fruit

Active ingredient : Leaves- Lupeol, β sitosterol, γ-sitosterol, Aegeline, Aegelenine, Rutin, Marmesinin, Glucoside Fruit- Psoralen, Aegelinol, Furanocoumarin, Marmelide, β -sitosterol

Pharmacological properties : Ferifuge, Antipyretic

Dosages : Take 5-6 leaves of bael, 1 stalk of neem, 1 piece of adrak, 4-5 black pepper; 5-6 leaves of tulsi, ½ cup coriander seeds, 1 clove flower, 10 gm jaggery. Grind all the ingredients. Boil in 2 glasses of water till one glass of decoction remain. Lastly add jaggery. Boil for one minute. Drink one glass liquid 3 times a day till the fever subsides.

Botanical Name : *Aristolochia indica*

Family : Aristolochiaceae

Common Name : Isharmul

Parts used : Fruit

Active ingredient : Aristolochic acid Aristolochic acid, D, Aristolamide, Aristolochine

Pharmacological properties : Ferifuge, Anti-periodic

Dosages : Rub the root on sandstone with a few drops of water to get 1 tsp paste. Apply the paste on the forehead twice a day.

Botanical Name : *Foreniculum vulgare*

Family : Apiaceae

Common Name : Sounf

Parts used : Seeds

Active ingredient : Flavonoids, Umbelliferone, Decursinal bergapten

Pharmacological properties : Ferifuge

Dosages : Grind one tsp Sounf, 10gm adrak, 25 gm jaggary. Add this in a glass of water, boil it for 5 minutes. Drink this liquid once a day till fever subsides.

Botanical Name : *Azadirachta indica*

Family : Meliaceae

Common Name : Neem

Parts used : Leaves

Active ingredient : Nimbin, Nimbolides, Nimocinolides, Isonimocinolides, Quercetin, β-sitosterol

Pharmacological properties : Anti-periodic, Anodyne, Anti-viral, Appetizer

Dosages : Extract juice from nem leaves and mix in a bowl of water. Take 1 cup daily till problem persists.

Botanical Name : *Allium cepa*

Family : Alliaceae

Common Name : Piyaz

Parts used : Bulb

Active ingredient : Organic sulphur compounds including trans-S-L-cysteine sulphoxide. Quercetin, Vanillic acid

Pharmacological properties : Febrifuge, Anti-pyretic, Anodyne

Dosages : Extract juice from one medium size Piyaz. Apply on the forehead to subside the fever.

Botanical Name : *Allium sativum*

Family : Alliaceae

Common Name : Lahsun

Parts used : Bulb

Active ingredient : Volatile oil, Alliin, Allinase, Vitamins-Thiamene, Riboflavin and Niacin

Pharmacological properties : Febrifuge, Anodyne, Tonic

Dosages : Take ½ bulb Lahsun and ½ tsp black pepper. Put them in one cup of water and boil for about 5 min. Strain it and add 1 tsp honey. Take one tsp once a day for 3 days.

Botanical Name : *Syzygium aromaticum*

Family : Myrtaceae

Common Name : Laung

Parts used : Bud

Active ingredient : Eugenin, Triterpene acid

Pharmacological properties : Febrifuge, Tonic

Dosages : Grind 2 cardamom and 2 cloves. Boil in 2 glasses of water till it reduces to ½ glass. Drink ½ glass once in a day till fever subsides.

Botanical Name : *Adhatoda vasica*

Family : Acanthaceae

Common Name : Adusa

Parts used : Leaves

Active ingredient : Vasicine Vasicol, Adhatonine, Vasicinone, Vasicinol, Vasicinolone

Pharmacological properties : Apperizer, Febrifuge

Dosages : Crush 100gm of Adusa leaves and soak in 1 cup hot water for 15 minutes and strain. Take ½ cup in the morning and in the evening for 3 days.

WEAKNESS

Botanical Name : *Curcuma longa*

Family : Zingiberaceae

Common Name : Haldi

Parts used : Rhizome

Active ingredient : Curcumin

Pharmacological properties : Stimulant, Thermogenic

Dosages : Boil 10gm of Haldi rhizome and 2gm Jira seeds in one glass of milk for 2-3 minutes. Drink one glass before going to sleep.

HYPERTENSION

Botanical Name : *Allium sativum*

Family : Alliaceae

Common Name : Lahsun

Parts used : Bulb

Active ingredient : Volatile oil, Alliin, Allinase, Vitamins-Thiamene, Riboflavin and Niacin

Pharmacological properties : Anti-chole sterol

Dosages : Peel the outer covering of 2-3 cloves of Lahsun. Take daily in an empty stomach.

Botanical Name : *Rauvolfia serpentine*

Family : Apocynaceae

Common Name : Sarpagandha

Parts used : Roots

Active ingredient : Reserpine, Deserpidine, Reseinamine

Pharmacological properties : Sedative

Dosages : Grind 10-15 roots and mix with juice of lime. Take one week ½ tsp twice a day, 2nd week ½ tsp once a day, 3rd week ½ tsp a day or alternate days, 4th week ½ tsp twice a week. 5th week ½ tsp only once a week.

— xxx —

Chapter 14

Jaundice

Botanical Name : *Aristolochia indica*

Family : Aristolochiaceae

Common Name : Isharmul

Parts used : Leaves

Active ingredient : Aristolochic acid, Aristolochine

Pharmacological properties : Digestive, Appetiser, Hepato protective

Dosages : Make paste with 5-6 leaves of Aristolochia indica, 5-6 leaves of Kalmegh and ½" root of Anantamul. Make pills of the size of red gram and dry in the shade. Take one pill daily for 10 days.

Botanical Name : *Andrographis paniculata*

Family : Acanthaceae

Common Name : Kalmegh

Parts used : Leaves/root

Active ingredient : Andrographo-lide, Homoandrographolide, Androgr-aphosterol

Pharmacological properties : Tonic, Hepatomegaly, Anti pyretic, Antibiotic

Dosages : Make paste with 5-6 leaves of Andrographis paniculata, 5-6 leaves of Isharmul and ½" root of Anantamul. Make pills of the size of red gram and dry in the shade. Take one daily for 10 days.

Botanical Name : *Aloe vera*

Family : Liliaceae

Common Name : Ghikunvar

Parts used : Leaves

Active ingredient : Barbaloin, Aloesin, Aloesone, Aloe-emodin

Pharmacological properties : Anthelmintic, Digestive, Spleenomegaly, Anti spasmodic, Regenerative

Dosages : Slit two medium size leaves to obtain the pulp and mix with a little Haldi and salt. Consume once a day for 15 days.

Botanical Name : *Adhatoda vasica*

Family : Acathanaceae

Common Name : Adusa

Parts used : Leaves

Active ingredient : Vasicine Vasicol, Adhatonine, Vasicinone, Vasicinol, Vasicinolone

Pharmacological properties : Anti emetic, Liver stimulant, Appetiser, Anti anaemic, Anti viral

Dosages : Mix 2 tablespoons of leaf juice with 1 teaspoon of honey. Take once a day for 10 days.

Botanical Name : *Allium sativum*

Family : Alliaceae

Common Name : Lahsun

Parts used : Bulb

Active ingredient : Volatile oil, Alliin, Allinase, Vitamins-Thiamene, Riboflavin and Niacin

Pharmacological properties : Liver stimulant, Digestive, Anti pyretic, Vermifuge, Anti inflammatory

Dosages : Boil one small bulb of Lahsun, 10-15 leaves of Bhringraja and 2 small pieces of Kattha in two glasses of water till water reduces to one glass. Drink twice daily for 11 days.

Botanical Name :*Hemidesmus indicus*

Family : Periplocaceae

Common Name : Kapuri

Parts used : Root

Active ingredient : b-sitosterol, a- amyrin, a-amyrin, Hexatriacontane Lupeol.

Pharmacological properties : Liver stimulant, Appetiser, Anti microbial, Anti pyretic, Anti emetic

Dosages : Make paste with 5-6 leaves of Kalmegh, 5-6 leaves of Isharmul and ½ roots of Kapuri. Make pills of the size of red gram with this paste and dry in the shade. Take one pill daily for 10 days.

Botanical Name : *Mentha arvensis*

Family : Lamiaceae

Common Name : Pudina

Parts used : Leaves

Active ingredient : Linalool, Menthofuran, Piperitone

Pharmacological properties : Anti spasmodic, Hepatomegaly, Spleeno-megaly, Anti emetic, Stimulant

Dosages : Boil 10g of leaves of Mentha arvensis in a glass of water, strain and cool. Add salt and black pepper for taste. Take once a day.

Botanical Name : *Phyllanthus niruri*

Family : Euphorbiaceae

Common Name : Bhuiamla

Parts used : Whole plant

Pharmacological properties :Phyllanthin-D, Phyllanthin, Hypophyllanthin

Dosages : Tonic, Digestion, Hepatopro-tective, Anti viral

1. Crush the whole plant part to make paste. Take once a day for 20-25 days on empty stomach.
2. Extract juice of 20-30ml leaves of Phyllanthus amarus and mix with one glass of buttermilk.

Botanical Name : *Piper nigrum*

Family : Piperaceae

Common Name : Kalimirch

Parts used : Seeds

Active ingredient : Piperine

Pharmacological properties : Digestive, Stimulant, Carminative, Anti inflammatory

Dosages : Powder 10gms each of Kalimirch, Adrak, Harra and Amla. Boil the powder in a little water in an iron vessel till the water evaporates for making tablets. Make tablets the size of ½ a bullet. Consume one tablet daily

Botanical Name : *Zingiber officinale*

Family : Zingiberaceae

Common Name : Adrak

Parts used : Rhizome

Active ingredient : Asparogine, Pipecolic acid

Pharmacological properties : Liver stimulant, Appetiser, Analgesic, Digestive,

Dosages : Grind ½" rhizome of Adrak and a small bulb of Piyaz. Add paste in two cups of water and boil to make one cup of decoction. Drink once a day for 3 days.

— xxx —

Chapter 15

Other Health Problem

WEEK NERVES

Botanical Name : *Piper betle*

Family : Piperaceae

Common Name : Paan

Parts used : Leaves

Active ingredient : sitosterols, Hentriacontane, Pentatriacontane, Stearic and Chavicol-Carvacrol, Eugenol, Chavicol, Allyl catechol, Cadinine

Pharmacological properties : Invigorating, Nervine tonic, Analgesic

Dosages : Extract juice of a few Paan leaves by crushing or grinding. Add one tsp of honey. Take one tsp twice a day.

HEADACHE

Botanical Name : *Piper betle*

Family : Piperaceae

Common Name : Paan

Parts used : Leaves

Active ingredient : β and γ-sitosterols, Hentriacontane, Pentatriacontane, Stearic and Chavicol-Carvacrol, Eugenol, Chavicol, Allyl catechol, Cadinine

Pharmacological properties : Invigorating, Nervine tonic, Analgesic

Dosages : Extract juice of a few Paan leaves by crushing or grinding. Apply juice on forehead.

ACIDITY

Botanical Name : *Asparagus racemosus*

Family : Asparagaceae

Common Name : Satavar

Parts used : Root

Active ingredient : Shatavarins, Saponins, Asparinins A and B, Asparoside A and B

Pharmacological properties : Stomachic, Appetizer

Dosages : Extract juice from 100gm roots. Take 6 tsp juice with sugar in the morning and evening for 7 days.

Botanical Name : *Tinospora cardifolia*

Family : Menispermaceae

Common Name : Giloy

Parts used : Stem

Active ingredient : Cardifolioside A & B, β-sitosterol, Tinosporin

Pharmacological properties : Digestive, Stomachic, Flatulent

Dosages : Extract juice from 100gm stems of Giloy. Take one tsp of juice thrice a day with honey for 7 days.

PEPTIC ULCER

Botanical Name : *Aegle marmelos*

Family : Rutaceae

Common Name : Bael

Parts used : Fruit, leaves

Active ingredient : Leaves- Lupeol, β sitosterol, γ-sitosterol, Aegeline, Aegelenine, Rutin, Marmesinin, Glucoside Fruit- Psoralen, Aegelinol, Furanocoumarin, Marmelide, β-sitosterol

Pharmacological properties : Anti-inflamma tory, Mucilagin

Dosages : Extract juice of fresh and ripe bel fruit by mashing on a sieve. Add milk and sugar to make it palatable in the form of sharbat. Drink a glass of sharbat daily for 2-3 months.

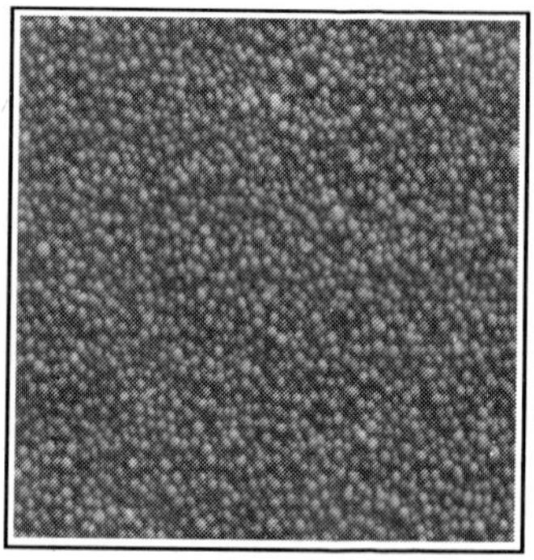

Botanical Name : *Brassica nigrun*

Family : Brassicaceae

Common Name : Kali rai

Parts used : Seeds

Active ingredient : Thioglucosides, Thioglucosinolates

Pharmacological properties : Anti pyretic, Anti inflammatory, Anti viral, Analgesic

Dosages : Grind one tablespoon seeds with one bulb of lahsun to make paste. Apply paste twice daily on the affected area.

Botanical Name : *Cajanus cajan*

Family : Fabaceae

Common Name : Arhar

Parts used : Tip of stem

Active ingredient : Cajanol

Pharmacological properties : Anti pyretic, Anti inflammatory, Anti viral, Analgesic

Dosages : Crush 2 tips of Cajanus cajan plant and extract the juice. Mix with ½ teaspoon sugar for consumption. Take three times a day with meals.

Botanical Name : *Datura stramonium*

Family : Solanaceae

Common Name : Datura

Parts used : Leaves

Active ingredient : Hyoscyamine, Hyoscine

Pharmacological properties : Anti pyretic, Anti inflammatory, Anti viral, Astringent, Narcotic

Dosages : Grind 2-3 leaves to get paste. Apply paste externally twice a day on the swollen glands

— xxx —

Chapter **17**

Nose Infection

Botanical Name : *Phyllanthus emblica*

Family : Euphorbiaceae

Common Name : Amla

Parts used : Fruit

Active ingredient : Vitamin C, Tannins, Gallic acid, Ellagic acid, Phyllemblic acid, Emblicol

Pharmacological properties : Anodyne, Anti-inflammatory, Alexeteric

Dosages : Extract juice from amla fruit. Pour two drops of juice into nose twice a day

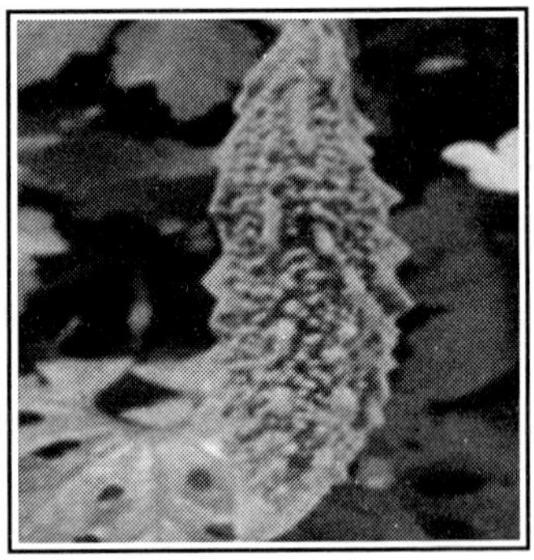

Botanical Name : *Mormordica charantia*

Family : Cucurbitaceae

Common Name : Karela

Parts used : Leaves

Active ingredient : Vines-tetracyclic, Triterpenes

Pharmacological properties : Anti-inflammatory, Vulnerary

Dosages : Grind two or three leaves to extract juice. Pour one or two drops into nose thrice a day.

Botanical Name : *Azadirachta indica*

Family : Meliaceae

Common Name : Neem

Parts used : Leaves

Active ingredient : Quercetin, β-sitosterol, Nimbin, Nimbolides, Nimocinolides, Isonimocinolides

Pharmacological properties : Anti-septic, Disinfectant, Demulcent

Dosages : Make fine paste of Neem leaves. Extract juice. Apply once a day on nose till symptoms persists

Botanical Name : *Mangifera indica*

Family : Anacardiaceae

Common Name : Aam

Parts used : Fruit

Active ingredient : Friedelin, β-sitosterol, Mangiferin, Catechin, Butin, Fisetin, Leucocyanidin, Quercetin extracted from various parts of the tree

Pharmacological properties : Anti-inflammatory, Vulnerary, Anti-bacterial

Dosages : Take unripe Aam and warm it on fire. Peel it off, squeeze the pulp to get juice. Add black salt. Drink 2-3 times a day for 2-3 days.

Botanical Name : *Ocimum gratissimum*

Family : Lamiaceae

Common Name : Rama Tulsi

Parts used : Leaves

Active ingredient : Ursolic acid, β-sitosterol, Cholesterol, Stigmasterol, Fatty acids

Pharmacological properties : Alexeteric, Demulcent

Dosages : Grind tusli leaves to extract juice. Put two drops in each nostril for 3-4 days

Botanical Name : *Acorus calams*

Family : Araceae

Common Name : Buch

Parts used : Roots

Active ingredient : Acorine, Choline

Pharmacological properties : Emetic, Alexipharmic

Dosages : Take 1 inch root and rub on a stone to make paste by adding water. Apply the paste on the skin immediately after the bite.

Botanical Name : *Acalypha indica*

Family : Euphorbiaceae

Common Name : Kuppi khokli

Parts used : Leaves

Active ingredient : Kaempferol, β-sitosterol , γ-sitosterol

Pharmacological properties : Emetic, Antidote

Dosages : Extract juice from 10-15 leaves. Instill few drops into nostrils.

Botanical Name : *Aristolochia indica*

Family : Aristolochiaceae

Common Name : Isharmul

Parts used : Roots

Active ingredient : Aristolochic acid, Aristochic acid D, Ariscolamide, Aristolochine

Pharmacological properties : Antidote, Astringent

Dosages : Crush 10gm roots to make fine paste. Apply the paste on the affected area.

Botanical Name : *Gymnenia sylvestra*

Family : Arclepiadaceae

Common Name : Gudmar

Parts used : Root

Active ingredient : Gymnemic acid, Hentriacontanol,

Pharmacological properties : Anodyne, Emetic, Astringent, Alexipharmic

Dosages : Grind 10gm roots to make powder. Prepare paste by adding water. Apply paste on the affected area.

Botanical Name : *Tamarindus indica*

Family : Caesalpiniaceae

Common Name : Imli

Parts used : Seeds

Active ingredient : Tartaric acid, Malic acid

Pharmacological properties : Anodyne, Anti-inflammatory, Astringent

Dosages : Grind 5-6 dry Imli seeds. Add little water to make paste. Apply on the affected area immediately after the bite.

INSECT BITE

Botanical Name : Azadirachta indica

Family : Meliaceae

Common Name : Neem

Parts used : Leaves

Active ingredient : Quercetin, β-sitosterol, Nimbin, Nimbolides, Nimocinolides Isonimocinolides

Pharmacological properties : Astringent, Emetic, Antidote, Anodyne

Dosages : Take 10-15 fresh Neem leaves. Eat 10-15 fresh leaves daily

Botanical Name : *Calotropis gigantean*

Family : Asclepiadaceae

Common Name : Madar

Parts used : Leaves

Active ingredient : Calotropin, Calatropagenin

Pharmacological properties : Astringent, Emetic

Dosages : Take out the milk by splitting the leaves. Immediately apply on the effect area the milk that oozes by breaking or splitting leaves

Botanical Name : *Eclipta prostrate*

Family : Asteraceae

Common Name : Bhringraja

Parts used : Roots

Active ingredient : Hentriacontanol, Heptacosanol Stigmasterol, Wedelolactone.

Pharmacological properties : Anodyne,Anti-inflammatory

Dosages : Make paste of 3-4 roots. Extracts juice of one small Piyaz by grinding. Mix the paste and the juice. Apply paste on the affected area

SCABIES

Botanical Name : *Acasia nilotica*

Family : Mimosaceae

Common Name : Babul

Parts used : Flower

Active ingredient : Polyphenolic compound

Pharmacological properties : Astringent, Emollient, Alexeteric, Nutritive, Liver tonic

Dosages : Grind 3-4 flowers and mix it in one tsp vinegar to make fine paste. Apply on the affected area for 7 days.

Botanical Name : *Azadiracta indica*

Family : Meliaceae

Common Name : Neem

Parts used : Stem

Active ingredient : Azadiractin, Nimbidin

Pharmacological properties : Astringent, Depurative, Demulcent, Anti-periodic, Alexeteric

Dosages : Grind 250gm dry leaves to make powder, take ½ tsb orally by adding sugar with hot water daily for 10 days.

Botanical Name : *Calotropis gigantean*

Family : Asclepiadaceae

Common Name : Madar

Parts used : Leaves

Active ingredient : Calotropin, Calatropagenin

Pharmacological properties : Astringent, Emetic

Dosages : Take out the milk by splitting the leaves. Immediately apply on the effect area the milk that oozes by breaking or splitting leaves

Botanical Name : *Eclipta prostrate*

Family : Asteraceae

Common Name : Bhringraja

Parts used : Roots

Active ingredient : Hentriacontanol, Heptacosanol Stigmasterol, Wedelolactone.

Pharmacological properties : Anodyne,Anti-inflammatory

Dosages : Make paste of 3-4 roots. Extracts juice of one small Piyaz by grinding. Mix the paste and the juice. Apply paste on the affected area

Botanical Name : *Terminalia chebula*

Family : Combretaceae

Common Name : Harra

Parts used : Seeds

Active ingredient : Chebulin, Chebulinic acid, Tannic acid

Pharmacological properties :Astringent, Anodyne, Anti-inflammatory

Dosages : Take 5gm seeds of Harra and grind in to make fine powder. Prepare paste by adding little water, apply this paste on the affected part immediately after the bite.

DOG BITE

Botanical Name : *Allium cepa*

Family : Alliaceae

Common Name : Piyaz

Parts used : Bulb

Active ingredient : Organic sulphur compounds including trans-S-L- cysteine sulphoxide. Quercetin, Vanillic acid

Pharmacological properties : Anti-bacterial, Refrigerant, Anodyne

Dosages : Take 5gm Piyaz, 2gm honey, 1gm wall nut,1gm salt. Make paste by adding water. Apply on the affected area immediately after the bite.

— xxx —

Chapter **19**

Skin Problems

BOILS AND WOUNDS

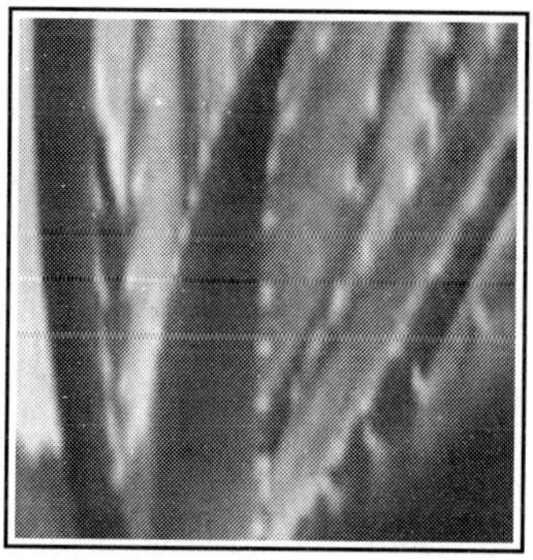

Botanical Name : *Aloe barbedensis*

Family : Liliaceae

Common Name : Ghikanvar

Parts used : Leaves

Active ingredient : Barbaloin, Aloesin, Aloesone, Aloe-emodin

Pharmacological properties : Alexeteric, Depurative, Cooling

Dosages : Split the leaves to get pulp. Apply this over the affected area twice a day till it cures.

Botanical Name : *Azadiracta indica*

Family : Meliaceae

Common Name : Neem

Parts used : Leaves & Bark

Active ingredient : Leaves- Quercetin, β-sitosterol, Nimbin, Nimbolides, Nimocinolides Isonimocinolides Bark- Azadiractin, Nimbidin, Nimbin, Nimbinin

Pharmacological properties : Antiseptic, Alexeteric, Vulnerary, Depurative

Dosages : Boil 22-25 leaves in water and mix it in cold water and take bath. Once a day till it cures.

Botanical Name : *Bacopa monnieri*

Family : Scrophulariaceae

Common Name : Brahmi

Parts used : Leaves

Active ingredient : Herpestine

Pharmacological properties : Astringent, Anodyne, Anti-inflammatory, Depurative

Dosages : Take 10-12 leaves and make fine paste by adding water. Apply on the boils twice a day for 4-5 days.

Botanical Name : *Colocasia esculenta*

Family : Araceae

Common Name : Aravi

Parts used : Leaves

Active ingredient :
3,4-diglucosilic benzaldehyde, Sabinene, Myrcene

Pharmacological properties : Demulcent,Anodyne

Dosages : Make ash by buring 2-3 dry Aravi leaves, mix it in mustard oil. Apply on the boil once a day for few days.

Botanical Name : *Curcuma longa*

Family : Zingiberaceae

Common Name : Haldi

Parts used : Rhizome

Active ingredient : Curcumin

Pharmacological properties : Anodyne, Anti-inflammatory, Vulnerary, Depurative

Dosages : Heat the ghee with Haldi powder in a pan to make fine paste. Apply this paste on the affected part twice a day for 3-4 days.

Botanical Name : *Lawsonia inermis*

Family : Lythraceae

Common Name : Henna

Parts used : Leaves

Active ingredient : Luteolin, β-sitosterol, Glucosides Lawsone

Pharmacological properties : Refrigerant, Anti-inflammatory, Vulnerary, Depurative

Dosages : Grind 15-20 leaves with curd to make fine paste. Apply this paste on the affected part twice a day for 3-4 days.

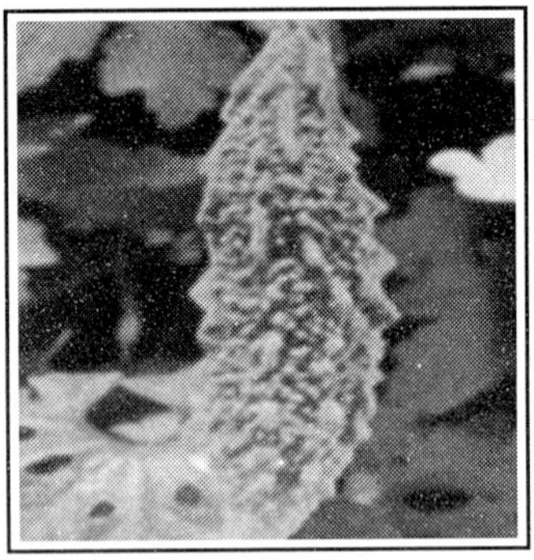

Botanical Name : *Momordica charantia*

Family : Cucurbitaceae

Common Name : Karela

Parts used : Fruit

Active ingredient : Cucurbitacin, Glycoside

Pharmacological properties : Anti-inflammatory, Depurative, Refrigerant, Vulnerary

Dosages :

1. Take 100ml juice of Karela and add 1tsp black pepper powder in it. Apply on the boils twice a day for 3-4 days.
2. Mix one cup of fresh Karela juice with 1tsp fresh lime juice. Drink one cup daily for 3-4 days in an empty stomach.

Botanical Name : *Piper betle*

Family : Piperaceae

Common Name : Paan

Parts used : Leaves

Active ingredient : β and γ-sitosterols, Hentriacontane, Pentatriacontane, Stearic and Chavicol-Carvacrol, Eugenol, Chavicol, Allyl catechol, Cadinine

Pharmacological properties : Antiseptic, Anti-inflammatory

Dosages : Extract juice of a few Paan leaves by crushing or grinding. Apply juice on wound and then wrap over wound with Paan leaf and bandage it.

Botanical Name : *Ricinus communis*

Family : Euphorbiaceae

Common Name : Arandi

Parts used : Seeds

Active ingredient : Triglycerides- ricinolein, Undecylenic acid

Pharmacological properties : Astringent, Anti, inflammatory Depurative

Dosages : Take 3-5 seeds of Erandi and prepare paste by adding water. Apply this paste twice a day for 3-4 days.

SCABIES

Botanical Name : *Acasia nilotica*

Family : Mimosaceae

Common Name : Babul

Parts used : Flower

Active ingredient : Polyphenolic compound

Pharmacological properties : Astringent, Emollient, Alexeteric, Nutritive, Liver tonic

Dosages : Grind 3-4 flowers and mix it in one tsp vinegar to make fine paste. Apply on the affected area for 7 days.

Botanical Name : *Azadiracta indica*

Family : Meliaceae

Common Name : Neem

Parts used : Stem

Active ingredient : Azadiractin, Nimbidin

Pharmacological properties : Astringent, Depurative, Demulcent, Anti-periodic, Alexeteric

Dosages : Grind 250gm dry leaves to make powder, take ½ tsb orally by adding sugar with hot water daily for 10 days.

Botanical Name : *Butea monosperma*

Family : Fabaceae

Common Name : Dhak

Parts used : Seeds

Active ingredient : Palasonin, β -sitosterol, γ-amyrin

Pharmacological properties : Astringent, Emollient, Anodyne, Anti-inflammatory, Depurative

Dosages : Grind 20gm seeds of Dhak with lemon juice to make a fine paste, Apply this paste regularly till the problem cures.

— xxx —

Chapter 20

Sore Throat

Botanical Name : *Ocimum gratissimum*

Family : Lamiaceae

Common Name : Rama Tulsi

Parts used : Leaves

Active ingredient : Ursolic acid, β-sitosterol, Cholesterol, Stigmasterol, Fatty acids

Pharmacological properties : Demulcent, Expectorant, Alexeteric

Dosages : Crush 6-7 Tusli leaves and extract juice. Take juice with honey 2-3 times a day

Botanical Name : *Terminalia chebula*

Family : Combretaceae

Common Name : Harra

Parts used : Fruit

Active ingredient : Chebulin, Chebulinic acid, Tannic acid

Pharmacological properties : Anti-inflammatory, Anti septic, Vulnerary, Febrifuge

Dosages : Roast Harra in ghee. Grind it into powder. Take one tea spoon with one glass of water twice a day for one week.

Botanical Name : *Bauhinia variegate*

Family : Caesalpiniaceae

Common Name : Kachnar

Parts used : Root and Bark

Active ingredient : Hentricontane, Octacosanol, Stigmasterol

Pharmacological properties : Anti-inflammatory, Vulnerary, Depurative

Dosages : Rub the kachnar root / bark and make a fine powder. Add half tea spoon to glass of water and gargle with

— xxx —

Chapter 21

Stomach Pain

Botanical Name : *Terminalia belerica*

Family : Combretaceae

Common Name : Bahera

Parts used : Fruit

Active ingredient : β -sitosterol, Galllic, Ellagic acids, Chebulagic acid and Bellaricanin

Pharmacological properties : Luxative, Astringent, Carminative, Anthelmintic

Dosages : Fry Harra and Bahera in ghee and grind to make powder. Take ¼ tsp of this powder with one glass of water twice a day for two days.

Botanical Name : *Aloe barbedensis*

Family : Liliaceae

Common Name : Ghikanvar

Parts used : Leaves

Active ingredient : Barbaloin, Aloesin, Aloesone, Aloe- emodin

Pharmacological properties : Carminative, Anthelmintic, Digestive, Stomachic, Aperient

Dosages : Make paste of the leaf of Ghikanvar. Take this paste once a day for two days.

Botanical Name : *Ocimum sactum*

Family : Lamiaceae

Common Name : Tulsi

Parts used : Leaves

Active ingredient : Ursolic acid, Luteolin

Pharmacological properties : Stomachic, Carminative, Vermifuge, Purgative, Anti-bacterial

Dosages : Grind 10-15 fresh leaves of Tulsi to make paste. Take this paste once a day.

Botanical Name : *Eclipta alba*

Family : Compositae

Common Name : Bhringaraja

Parts used : Leaves

Active ingredient :
Stigmasterol, Wedelolactone, Desmethylwelo-Lactone, Terthienylmethanol

Pharmacological properties : Carminative

Dosages : Grind 15-20 fresh leaves of Bhringaraja and Mandukaparni to make paste. Take this paste twice a day with meal for 1-2 days.

Botanical Name : *Centella asiatica*

Family : Hydrocotylaceae

Common Name : Mandukaparni

Parts used : Leaves

Active ingredient :
Brahmoside, Braminoside Brahmic acid, Amino acids

Pharmacological properties : Tonic, Cooling, Luxative, Appetizer, Carminative

Dosages : Grind 15-20 fresh leaves of Bhringaraja and Mandukaparni to make paste. Take this paste twice a day with meal for 1-2 days.

Botanical Name : *Ocimum basillicum*

Family : Lamiaceae

Common Name : Tulsi

Parts used : Leaves

Active ingredient : Eugenol, Limonene, Geraniol, Citronellol, α-terpineol, Ocimene

Pharmacological properties : Anti-flatulent, Stomachic, Anthelmintic, Stimulant, Demulcent

Dosages : Boil 5-6 fresh leaves each of Tulsi and Pudina and ½ tbp of Ajowan seeds in sufficient water till quantity reduces to half. Take 1-2 tsp of decoction when pain arises.

Botanical Name : *Casia auriculata*

Family : Caesalpiniaceae

Common Name : Tarwar

Parts used : Flower

Active ingredient : β -sitosterol and Kaempferol,

Pharmacological properties : Anthelmintic, Stimulant, Astringent,

Dosages : Boil bark of Tarwar in sufficient water till the quantity reduces to half. Take one tsp of decoction twice a day for getting immediate relief.

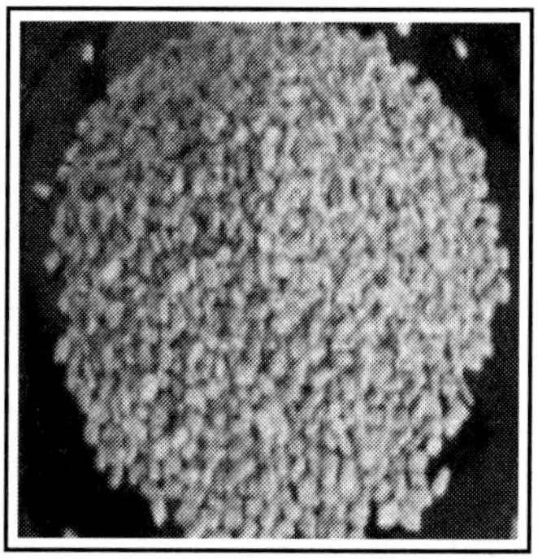

Botanical Name : *Trigonella foenum graceum*

Family : Fabaceae

Common Name : Methi

Parts used : Seeds

Active ingredient : Sapogenins, Saponins, Fenugreekine

Pharmacological properties : Anthelmintic, Antispapsmodic, Carminative, Antiseptic

Dosages : Mix one tsp of Fenugreek seeds with a pinch of Hing. Take twice a day with warm water.

Botanical Name : *Ficus bangalensis*

Family : Moraceae

Common Name : Bargad

Parts used : Leaves

Active ingredient : β-sitosterol, Rutin

Pharmacological properties : Anodyne, Astringent, Cooling, Antiseptic

Dosages : Mix ash of 2-3 leaves of Banyan tree in a glass of water. Drink one glass once a day.

Botanical Name : *Murraya koenigii*

Family : Rutaceae

Common Name : Mithaneam

Parts used : Leaves

Active ingredient : Coumarin, Glucoside, Scopolin

Pharmacological properties : Supportive Stimulant, Anti-flatulent, Tonic

Dosages : Grind 15-20 Curry leaves to make to make paste. Mix it with one cup of butter milk . Drink it in empty stomach in the morning.

Botanical Name : *Embelica ribes*

Family : Myrsinaceae

Common Name : Vidanga

Parts used : Seeds

Active ingredient : Quinonesembelin, Repanone, Homoembelin Homorapanone & Vilangin

Pharmacological properties : Anthelmintic, Anti-flatuent

Dosages : Boil 15-20 seeds in one glass of milk or water and strain it. Drink half cup thrice a day.

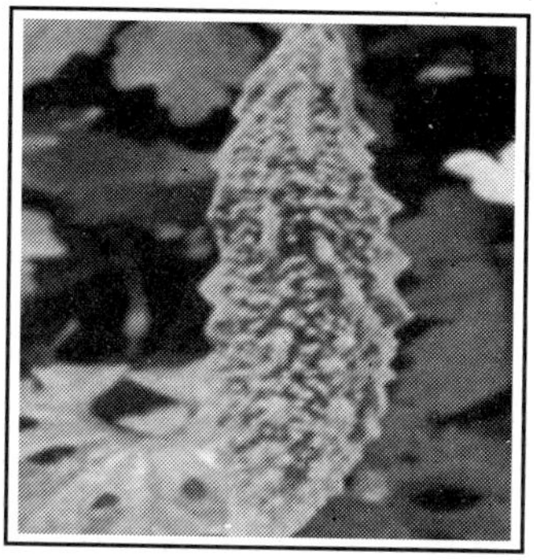

Botanical Name : *Momordica charantia*

Family : Cucurbitaceae

Common Name : Karela

Parts used : Fruit

Active ingredient : Cucurbitacin, Glycosides

Pharmacological properties : Carminative, Tonic, Stomachic, Anthelmintic, Astringent, Luxative

Dosages : Ground one Karela and mix pulp in half cup of milk and pinch of Haldi. Take twice a day till condition improves.

Botanical Name : *Calotropis gigantean*

Family : Asclepiadaceae

Common Name : Madar

Parts used : Leaves

Active ingredient : Calotropin, Calatropagenin Anthelmintic,

Pharmacological properties : Luxative, Astringent, Purgative

Dosages : Grind 3-4 leaves of Madar and mix well with Jaggery. Take once a day.

— xxx —

Chapter 22

Tonsillitis

Botanical Name : *Acacia nilotica*

Family : Mimosaceae

Common Name : Babul

Parts used : Bark

Active ingredient : Polyphenolic compounds

Pharmacological properties : Alexeteric, Emetic, Vulnerary, Nutritive

Dosages : Boil 5gms bark in water. Grind the boiled bark to a fine paste and add 10gms powdered sugar. Give half of the total quantity twice a day for 2-3 days

Botanical Name : *Allium sativum*

Family : Alliaceae

Common Name : Lahsun

Parts used : Bulb

Active ingredient : Volatile oil, Alliin, Allinase, Vitamins-Thiamene, Riboflavin and Niacin

Pharmacological properties : Anodyne, Alexeteric, Anti-bacterial, Expectorant, Tonic

Dosages : Peel 50gms of Lahsun pods and crush them. Extract juice and add half teaspoon of honey. Mix well. Now consume the quantity once in a day for 2-3 days.

Botanical Name : *Brassica juncea*

Family : Brassicaceae

Common Name : Rai

Parts used : Seeds

Active ingredient : Brassicasterol, Thiocynates, Glucosinolates

Pharmacological properties : Thermogenic, Anti inflammatory, Anodyne, Tonic, Digestive

Dosages :

1. Boil 10gms mustard leaves in 100ml of water for 2-3 min and add pinch of salt. Strain it. Now drink warm liquid once a day.
2. Massage warm mustard oil on the neck once a day message for 5 days.

Botanical Name : *Sapindus laurifolius*

Family : Sapindaceae

Common Name : Ritha

Parts used : Fruit

Active ingredient : Saponin, Sapindic acid, Oleanolate

Pharmacological properties : Expectorant, Thermogenic

Dosages : Extract juice of two pieces of fresh soap nut fruit. Put two drops in each ear once a day.

Botanical Name : *Syzygium aromaticum*

Family : Myrtaceae

Common Name : Laung

Parts used : Flowers

Active ingredient : Eugenin, Triterpine acids, Eugenol, Caryophyllene, Eugenyl acetate, Chavicol

Pharmacological properties : Expectorant, Appetizer, Emollient, Anti-bacterial, Tonic

Dosages : Grind one Laung with 8-10 tulsi leaves, one pepper and four cloves of Lahsun. Make a thick paste and divide into seven portions. Take one portion once a day for one week.

Botanical Name : *Capsicum annum*

Family : Solanaecae

Common Name : Lal mirch

Parts used : Fruit

Active ingredient : Capsaicin, Homocapsaicin, Vanillylamide Decyate.

Pharmacological properties : Stimulant, Expectorant, Thermogenic, Sialogogue

Dosages : Mix half the spoon red chilly powder with crushed garlic (5-6 cloves). Eat with dry chapati.

Botanical Name : *Curcuma longa*

Family : Zingiberaceae

Common Name : Haldi

Parts used : Root

Active ingredient : Curcumin

Pharmacological properties : Thermogenic, Anti inflammatory, Anti septic, Anodyne, Emollient

Dosages : Grind the root to make a fine powder. Take one teaspoon of powder with water twice for 3-4 days.

Botanical Name : *Eucalyptus globulus*

Family : Myrtaceae

Common Name : Eucalyptus

Parts used : Leaves

Active ingredient : Camphene, Sabinene, α and β -thujone, Verbenone, Flavonoides

Pharmacological properties : Febrifuge, Expectorant, Antiseptic, Anodyne

Dosages : Burns 7-8 leaves and inhales the fume every night before going to sleep. Inhale fumes for 5-6 days.

Botanical Name : *Daucus carota*

Family : Apiaceae

Common Name : Gajar

Parts used : Root

Active ingredient : Essential oil contains pinene, Camphene, Myrcene, Car-3-ene, Limonene, β-bisabolene, β-caryophellene, β-selinene, Carotol, Daucol

Pharmacological properties : Thermogenic, Expectorant, Vulnerary

Dosages : Eat 6 carrots 3-4 times a day

Botanical Name : *Ricinus communis*

Family : Euphorbiaceae

Common Name : Arandi

Parts used : Leaves

Active ingredient : Amino acids

Pharmacological properties : Thermogenic, Emollient, Anti inflammatory

Dosages : Wash 5-6 leaves of Erandi and crush to extract juice. Take juice with one glass of hot cow's milk every night before going off to sleep for 3 days.

— xxx —

Chapter 23

Vomiting

Botanical Name : *Mentha arvensis*

Family : Lamiaceae

Common Name : Pudina

Parts used : Leaves

Active ingredient : Menthol, Menthane, Methyl acetate

Pharmacological properties : Anthelmintic, Antiseptic, Refrigerant, Stomachic, Tonic

Dosages : Grind 15-20 fresh leaves of Pudina and a small piece of adrak to extract juice. Take this juice with one glass of water thrice a day.

Botanical Name : *Ocimum sanctum*

Family : Lamiaceae

Common Name : Tulsi

Parts used : Leaves

Active ingredient : Ursolic acid, Luteolin

Pharmacological properties : Stomachic, Digestive, Purgative, Anti-poisonous

Dosages : Extract juice of 15-20 fresh leaves of Tulsi and add one tsp of honey. Take one tsp once a day till condition improves.

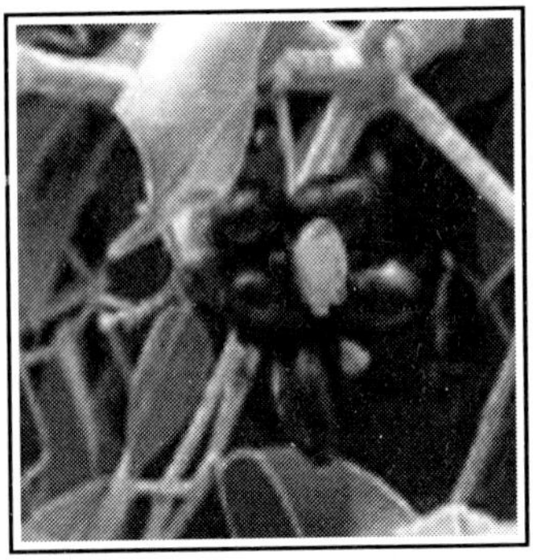

Botanical Name : *Syzygium cumini*

Family : Myrtaceae

Common Name : Jamun

Parts used : Leaves

Active ingredient : Oleanolic acid, Crategolic acid

Pharmacological properties : Stomachic, Refrigerant, Anti-bacterial, Cooling

Dosages : Chew 1-2 fresh tender leaves once a day.

Botanical Name : *Aegle marmelos*

Family : Rutaceae

Common Name : Bael

Parts used : Fruit

Active ingredient : Psoralen, Aegelinol, Furanocoumarin, Marmelide, β-sitosterol

Pharmacological properties : Appetiser, Binding, Tonic, Restorative, Astringent, Luxative

Dosages : Make paste of pulp of one fruit of Bael and unripe seeds of mango in eqal proportion. Take 6-8 tsp of this paste twice till condition improves.

Botanical Name : *Ficus religiosa*

Family : Moraceae

Common Name : Pipal

Parts used : Bark

Active ingredient : β-sitosterol and its glucosides

Pharmacological properties : Astringent, Cooling, Anti-bacterial

Dosages : Make ash of bark of Pipal after drying in shade. Mix one tsp of ash with a glass of water and strain. Drink one glass when vomiting occurs.

Botanical Name : *Ficus bangalensis*

Family : Moraceae

Common Name : Bargad

Parts used : Roots

Active ingredient : Leucoanthocyanin Leuco-anthocyanidin

Pharmacological properties : Anodyne, Astringent, Cooling, Antiseptic

Dosages : Crush roots of Bargad tree to extract juice. Take 4-5 tsp juice twice a day.

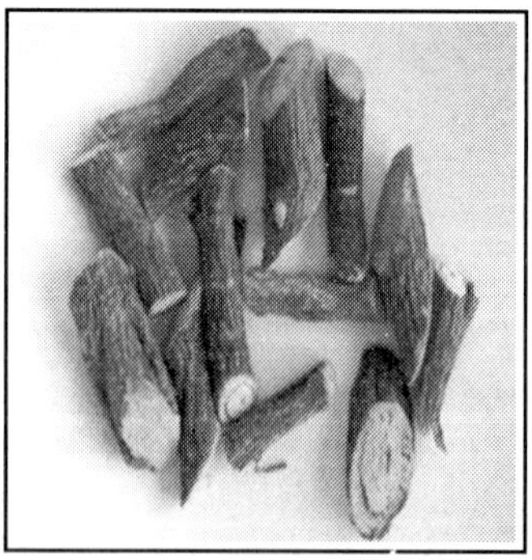

Botanical Name : *Glycyrrhiza glabra*

Family : Fabaceae

Common Name : Mulathee

Parts used : Root

Active ingredient : Glycyrrhizin, Glycyrrhizic acid

Pharmacological properties : Suportive Demulcent, Anti-bacterial

Dosages : Make paste of ½ inch root of Mulathee with few drops of water by rubbing on stone. Mix this paste with one tsp of honey. Take once a day.

Botanical Name : *Anethum sowa*

Family : Apiaceae

Common Name : Sowa

Parts used : Seeds

Active ingredient : Carvone, Limonene, Dihydrocarvone, Dillapeole

Pharmacological properties : Demulcent, Astringent, Carminative

Dosages : Grind one tsp of Sowa seeds to make powder and mix in a glass of water. Drink one glass once a day.

Botanical Name : *Coriandum sativum*

Family : Apiaceae

Common Name : Dhania

Parts used : Seeds

Active ingredient : Linalool, Camphene, Geraniol

Pharmacological properties : Stomachic, Stimulant, Anthelmintic, Carminative, Antispasmodic

Dosages : Grind one tsp of Dhania to make powder. Boil in a glass of water or milk for about 5 min and add one tsp of sugar. Drink once a day.

Botanical Name : *Citrus auratifolia*

Family : Rutaceae

Common Name : Kagzi Nimbu

Parts used : Fruit

Active ingredient : An aldehyde octanal in lime oil, Neryl, Geranyl acetates.

Pharmacological properties : Anthelmintic, Appetizer, Refrigerant

Dosages : Grind one fruit of lime, one tsp seeds of Pepper and one rhizome of Sonth to make powder. Mix with limejuice. Take thrice a day.

— xxx —

Chapter 24

Plants Medicine

Approaches to Plant – Derived Drug Development

Brief history

Fossils of plant date back as early as 3.2 billion years ago. These plans provided the foundation upon which animal life and later, human life were based on. They provide bodybuilding food and calories as well as vitamins essential for metabolic regulation. Plants also yield active principles employed as medicines.

Finding healing powers in plants is an ancient idea. Hundreds, if not thousands, of indigenous plants have been used by people on all continents as poultices and in dating back to prehistory. There is evidence of Neanderthals, living 60 000 years ago in present-day Iraq, using hollyhock (*Aocea rosea* L.), which is still in ethnomedicinal use around the world today. The Bible offers descriptions of at least 30 healing plants of which frankincense (*Boswellia sacra* L.)

and myrrh (*Commiphora myrrha* L.) were employed as mouthwashes due to their reported antiseptic properties.

The fall of ancient civilizations resulted in the destruction or loss of much of the documentation of plant pharmaceuticals by many cultures continued in the excavation of the older works as well as building upon them. Native Americans were reported to have used 1625 species of plants as food while 2564 found use as drugs, while the Europeans started turning towards botanicals when treatment in the 1800s become dangerous and ineffective. Today some 1500 species of medicinal and aromatic plants are widely used in Albania, Bulgaria, Croatia, France, Germany, Hungary, Spain, Turkey and the United Kingdom.

Origin of pharmacology

Natural products have been used to elucidate physiological processes and even define them, hence the naming of 'nicotinic' and 'muscarinic' receptors and even more recently 'endorphins' from 'endogenous morphines'. Natural products are the basis of many standard drugs used in modern medicine and are so widely used that even some members of the medicinal profession are not aware of their plant origin. Some of the newer pharmacological tools such as colforsin (an adenyl cyclase stimulator), ginkgolide B {a specific platelet activating factor (PAF) antagonist} and phorbol esters that activate proteinkinase C, are at the forefront of biochemical research and are obtainable only from plant sources. Although medicinal plants may not always lead to the discovery of novel compounds which may be employed in the treatment or cure of disease, plants may give valuable insight into the pathology of diseased conditions or the disturbed human mind. Hallucinogenic plants have been thought to transport the mind to realms of ethereal wonder and some were even considered to be gods. It is only recently, the past twenty years, that modern westernized societies have realized the significance of these plants in shaping the history of primitive and even advanced cultures. Some of these plants contain chemicals capable of inducing visual, auditory, tactile, olfactory and gustatory hallucinations or causing artificial psychoses and the question is whether a thorough understanding of the chemical composition of these drugs may lead to discovery of new drugs for the treatment of psychiatric conditions. As a result of the complexity of the human brain and central nervous system, psychiatry has not developed as rapidly as other fields of medicine, mainly due to the lack of adequate tools, therefore these drugs may provide the necessary pharmacological tools for the discovery of more appropriate and effective drugs.

Although the first chemical substance to be isolated from plants was benzoic acid in 1560, the search for useful drugs of known structure did not begin until 1804 when morphine was separated form *Papaver somniferum* L. (Opium). Since then many drugs from higher plants have been discovered but less than 100 of defined structure are in common use today. Less than half of these are accepted as useful drugs in industrialized countries [Farnsworth, 1984].

Secondary plant compounds

Living organisms, whether they are plants, animals or microbes, are all made up of basically the same chemical components. Comparative biochemistry has shown that minor or major modifications of chemical processes have been evolved by different organisms to suit their particular requirements. Chemical compounds have been extensively used in plant systematics and two major categories of systematically useful chemical compounds can be recognized: secondary metabolites, which perform non-essential functions in the plant, and the information containing proteins, DNA and RNA.

Most secondary compounds function in defence against predators and pathogens, as allelpathic agents or attractants in pollination and seed dispersion. The major categories are discussed briefly below.

Active constituents

Terpenoids are a large and structurally diverse group of secondary compounds that are important is numerous biotic interactions. They are widely distributed and many have primary physiological functions, either forming part of membrane-bound steroids, carotenoid pigments, side-chain of chlorophyll and the hormones giberellic acid and abcisic acid. Volatile monoterpenoids are the major components of essential oils and often function as floral odour glands. More information regarding these compounds are discussed in chapter.

The active ingredients are the main effective compounds of medicinal plants. The quantification of active in gradients in the first step of standardization of crude drugs. A number of chemical test are available for quantification of active ingredients. Some of the main chemical group (organic compounds) of active ingredients are:-

1. Alkaloids

Cinchona alkaloids present in the bark of *Conchona spp.* Have quinine as their main constituent, known since 1630 for its antimalarial properties. All alkaloids contain nitrogen, which frequently forms part of a heterocylic ring and make them basic in nature. Widespread distribution in all parts of plants has stimulated searches for a function of these compounds in the general metabolism of plants. These compounds offer protection against predators, act as growth regulators, maintain ionic balance, act as a nitrogen reserve and possibly serve as nitrogen excretion products.

2. Flavonids

These represent a widespread group of water-soluble phenolic derivatives, many of which are brightly coloured. At least nine classes of flavonoids are recognized and are of great interest in phytochemistry. They probably function in defence against herbivores and in regulation of auxin transport. Attraction of insects and birds also play an important role in seed dispersal and pollination.

3. Saponines

They are often strong expectorants, and stay also aid in the absorption of nutrients.

4. Cardiac Glycosides

They are diuretic. They help to transfer fluids from the tissues and circulatory system to the urinary tract, thereby lowering blood pressure.

5. Mucilage

It is made up of polysaccharides (large sugar molecules) that soak up water, producing a sticky jelly-like mass. Mucilage lines the mucous membranes of the digestive tract, protecting against irritation, acidity and inflammation.

6. Phenols

This group of compounds includes salicylic acid. Phenols are antiseptic and reduce inflammation.

7. Tannin

These are produced to a greater or lesser degree by the plants. They drawn the tissues closer together and improve their resistance to infection.

8. Coumarins

Coumarins of different kinds are found in many plant species and have widely divergent actins. It is a powerful smooth muscle relaxant.

9. Anthraquinones

Anthraquionones have an irritant laxative effect on the large intestine.

10. Anthocyanins

Help to keep the blood vessels healthy.

11. Glucosilinates

These compound are found exclusively some medicinal plants belong to mustard family. Glucosilinates have an irritant effect on the skin, causing inflammation and blistering. Glucosilinates also help to reduce thyroid function.

12. Volatile oils

It is extracted from plants to produce essential oils. Some volatile oils contain sesquiterpenes. These constituents have an anti-inflammatory effect. And many of them being strongly antiseptic.

13. Cyanogenic glycosides

Though these glycosides are based on cyanide, a very potent poison. They have a helpful sedative and relaxant effect.

14. Vitamins

Some plants contain significant levels of vitamins. Most other medicinal plants contain at least some vitamins.

15. Bitters

Bitterness itself stimulates secretions by the salivary glands and digestive organs. Such secretions can dramatically improve the appetite and strengthen the overall function of the digestive system.

16. Minerals

A number of medicinal plants are rich in minerals it help to maintain high levels of potassium.

Plants As A Source of Antibiotics

Plants have developed an arsenal of weapons to survive attacks by microbial invasion. These include both physical barriers as well as chemical ones. i.e. the presence or accumulation of antimicrobial metabolites. These are either preformed in the plant (prohibitins) or induced after infection, the so-called phytoalexins. Since phytoalexins can also be induced by abiotic factors such as UV irradiation, they have been defined as 'antibiotics formed in plants via a metabolic sequence induced either biotically or in response to chemical or environmental factors'.

When an infection or damage to a plant takes place, a number of processes are activated and some of the compounds produced become activated immediately whereas phytoalexins take two to three days to produce. Sometimes it is difficult to determine whether the compounds are phytoalesons or prohibitions and moreover, the same compound may be a preformed antimicrobial substance in one species and a phytoalexin in another. The chemical class in which these substances can be found varies greatly, which makes the isolation thereof even more intriguing.

Since the advent of antibiotics in the 1950s, the use of plant derivatives as antimicrobials has been virtually non-existent but that pace is rapidly on the increase as we begin to realise the need for new and effective treatments. The worldwide spending on finding new anti-infective agents is expected to increase 60% from 1993 and plant sources are especially being investigated.

Literally thousands of phytochemicals with inhibitory effects on microorganisms have been found to be active in vitro. One may argue that these compounds have not been tested in vivo and therefore activity cannot be claimed but one must take into consideration that many, if not all, of these plants have been used for centuries by various cultures in the treatment of disease. Another argument could possibly be that at very high concentrations, any compound is likely to inhibit the growth of microorganisms. Firstly if this is the case, the high concentrations required would no doubt have serious side-effects on the patient unfortunate enough to contract an illness and secondly, these compounds are compared with those of standard antibiotics already available on the market. This means that the concentrations used must compare favourably to those that have already passed the test.

Asiaticoside, and antimicrobial compound isolated from *Centella asiatica* (used traditionally in skin disease and leprosy), has been studied in normal as well as delayed-type wound healing. The results indicated significant wound healing in both models. Another compound, cryptolepine, isolated from *Cryptolepis sanguinolenta* and active against *Camphlobacter* species, has been used traditionally in Guinea Bissau in the treatment of hepatitis and in Ghana for the treatment of urinary and upper respiratory tract infections and malaria.

Preliminary screening tests performed by Noristan scientists on a number of plants species resulted in the finding that *Combratum erythrophyllum* had antimicrobial activity but due to changes in company policy no further work was performed. A comprehensive study into this family and specie was therefore initiated to determine whether there was activity and whether these compounds could be isolated.

Good Agricultural Practices (GAP)

The survival of natural plant varieties can only be ensured, if industrial needs of their raw materials and derivations is met through other means of productions, at competitive prices.

Knowledge of genetics and plant physiology in medicinal and aromatic plants is unfortunately not as developed as that of food crops. Our knowledge of the biosynthetic pathways that lead to generation of physiologically important plant constituents which make these crops economically valuable, is still in its infancy. Commercial cultivation of medicinal and aromatic plants species presupposes strong and continuing research. The contemporary literature in agrotechnolgoy of several medicinal and aromatic plants demonstrates the fact that plant varieties with an abundance of desired constituents can be reproduced, and improved upon under cultivation even in an alien habitat. Knowledge of breeding behaviour, and genetic architecture of a species helps to evolve new cultivars with reduced crop duration in the field, enhanced fertilizer responsiveness, improved yield of the desired secondary metabolite and reduction of any undesirable constituents [Franz 1993].

The cultivation of medicinal and aromatic plants is dictated by the requirements of the pharmaceutical and the fragrance industries which are the main users of the raw material. Bulk production of these plants species is regulated by international trade except in instances where there is a large internal demand. Adoption of GAP will help in maintaining the parameters required for the industrial use of plants. In the Region, several countries, notably China, India, Thailand, Korea, Sri Lanka, Vietnam, Indonesia, Nepal and the Philippines are known to have crop cultivation of medicinal and aromatic plants species for internal use as well as for export.

Improvements in Processing Technology

After harvest, some materials have to be dried in the sun, others in the shade, still other for protracted periods

quantity of water has been poured on to it at room temperature. The extract is then strained and made up to the prescribed weight with rinsings.

- *Tinctures* : Tinctures are extracts from drug plants prepared with ethanol of varying concentration, ether or mixtures of these, perhaps with certain additives, in such a way that one part of drug is extracted with more than two parts, but at most ten parts, of extraction liquid.

- *Preservation* : Medicinal plants can be preserved in a number of ways, the most common and simple, being air or oven drying. A warm, dry place, for example an airing cupboard, is ideal use plain paper for drying herbs, never printed newspaper. Dried herbs can be stored in a dark glass jar or a brown paper bag.

- *Swelling of the Plant Material* : To circumvent any problems, in all industrial procedures the raw material is pre-treated with solvent outside the extractor prior to charging the latter. This prevents sudden bulk volume changes, the causing of "channeling" during extraction, and facilitates the breaking of the cell walls to release the extractables.

- *Diffusion, pH, Particle Size and Temperature* : To facilitate extraction, the solvent should diffuse inside the cell and the substance must be sufficiently soluble in the solvent. The speed with which an equilibrium is attained between solute and solvent is dependent on it.

- *Choice of Solvent* : The ideal solvent for complete extraction is one that is the most selective, has the best capacity for extraction, and is compatible with the properties of the material to be extracted. These parameters must be pre-determined experimentally. In the industrial context, the factors of economy and availability come to the fore. Alcohol is widely used but because of its great extractive power it is often the least selective, in that it extracts all soluble constituents. Alcohol in various ratios are used to minimize selectivity. The ideal alcohol ratio for woody or bark material is 75%. For leafy material it is often less than 50% thus avoiding extraction of the chlorophylls which make purification problematical.

- *Use of Chemical Properties* : Some materials such as alkaloids being soluble in acids, their extraction is facilitated by the pH in the acid region. A number of alkaloids can be extracted easily with hydro carbons after they have been released from combination with organic acids by grinding with alkali. Thus cones sine is first ground with moist calcium oxide and extracted with chloroform. Another large number of alkaloids can be extracted directly with aqueous acids, organic or inorganic, and the alkalised extracts counter-extracted with hydrogcarbons or other apolar solvents. The extraction procedure has to be tailored to the given situation.

USES OF MEDICINAL PLANTS

Medicinal plants are used in the following ways :

As raw materials for extraction of active constituents in a pure form, e.g. alkaloids (quinine, quinidine) from cinchona bark, emetine from ipecacuanha root, glycosides from digaitalis leaves, sennosides from senna leaves.

As precursors for synthetic vitamins or steroids.

As preparation for herbal and indigenous medicines.

For direct sale when dried, cleaned, ground and packed, for pharmacies. Products such as ginseng, valerian and liquorice roots are part of the herbal and health food market as well as the food, flavour, fragrance and cosmetic industries. Less than 20% are used directly as pharmaceuticals. Examples :

Liquorice	-	confectionery and tobacco
Papain	-	meat tenderizer
Quinine	-	soft drink tonic
Cinchone	-	wine flavour

A large quantity of medicinal plant material is used in the preparation of herbal and medicinal teas, e.g. eamomile. These herbal and food uses are of great importance, also to the exporters from developing countries. Hundreds of medicinal plants are items of commerce, however, relatively small quantities are used in formulated 'herbal' remedies.

Extraction and Utilization of Aroma principle

Plants owe their fragrance to the presence of traces of essential oils in different part. Numerous fragrance materials are present in roots stems, barks, leaves, flowers, fruits and heartwoods. Gums, balsams and oleoresins are also valuable raw materials for perfumes by virtue of their tenacious but sot odour. Several processes like hydrodistillation, stem distillation, hydrodiffusion, enfleurage, maceratin, expression and solvent extraction are available for the extraction of aroma principles. Application of these processes, either singly or in combination, depends upon the nature of the material and of the essential oil or absolute intended to be recovered.

Distillation of essential oils

Essential oils represent the "essential aroma" of the plant from which they are obtained. The majority of essential oils are produced by the process of steam distillation. For this the plant material must be finely divided or comminuted.

Essential oils occur in many different parts of plants, e.g. roots (vetiver), bark (cinnamon), heartwood (sandalwood), leaves (bay), herb (peppermint), seeds (nutmeg), flowers (canange and jasmine).

The essential oil of a plant consists of many compounds which generally boil at between 150-300°C. If attempts are made to remove these compounds by dry distillation many will decompose and the oil will be ruined. However, the compounds are steam volatile and can be distilled out of the vegetable materials at around 100°C.

When plant materials are steam-distilled chemical changes inevitably occur and the oil obtained will not have an aroma identical to that of the original plant material. Post-processing oxidation of the oils is a common problem to be avoided by exclusion of air, trace metals and sunlight.

Types of Distillation

There are three basic type of essential oil distillation :

1. Hydrodistillation
2. Wet steam
3. Dry steam

Hydrodistillation : The charge is usually comminuted and immersed in boiling water. Stills are the simplest type used by small producers of essential oils. The disadvantages of an open fire are that the heat is difficult to control and the rate of distillation is variable. Local overheating and "burning" of the charge can lead to poor quality oil. Improved distillation control can be obtained by using steam from a separate boiler, into a steam jacket or through a closed steam coil, to heat the still. This system requires the heating of large quantity of water, adding to costs, high fuel wool consumption and time needed for each distillation. However, it is necessary for certain flower distillations, e.g. rose and ylang. It is also necessary for the efficient distillation of certain woody materials, e.g. sandalwood and cinnamon bark.

Wet steam distillation

This is an improved method. The still contains a grid which keeps the plant material above the water level. The water is boiled below the charge and "wet" steam passes through the plant material. If an open fire is used the plant material is protected from direct heat. The still is heated by a steam jacket. It is important in both water/steam and steam distillation that the still is packed evenly and not too tightly so that steam can extract from the complete charge efficiently. Over packing of the still can cause the steam to force holes through the charge and leave parts unextracted.

Dry steam distillation

This is a more advanced type of distillation, by direct steam provided from a separate boiler. The still contains a grid plate under which an open steam 'sparge' pipe is fitted. The advantages of this type of "dry" steam distillation are that it is rapid, charging and emptying the still is much faster, and energy consumption is lower. The rapid distillation is also less likely to damage those oils which contain reactive compounds, e.g. esters.

All stills should be insulated to reduce heat losses. Their design and construction should also easy loading and unloading.

Steam Distillation of Essential Oils

In processing, steam forms a constant boiling mixture with oils which collectively boils at a temperature below that of steam. Hence the passage of steam through a bed of plant material causes the volatiles to distil over with the steam. When the condensate cools, the oils, relatively insoluble in water, separate from the aqueous phase, forming two layer, which then can easily be separated. The equipment for the distillation can vary with the method used. Generally, steam can be generated *in situ*, with the bottom of the still serving as a boiler. In better designs there is a separate steam source, in the form of a satellite boiler.

Essential oils have established quality parameters and the production must also coupled with a facility for instrumental analysis of the product. Here, too the methodology has many variation and must be suited to the particular plant species and its constituent oil. For example, factors such as the following are important in that context:

Optimum size of the still (Load capacity)

Steam pressure

Density of packing of the plant material in the still body

Duration of distillation

Rate of steam injection

All these factors have to be determined experimentally for obtainment of the best possible yield and a product of approved quality. Application at industrial scale must be preceded by appropriate trials (Wijeskera 1993).

Condensers and Separators

The steam with essential oil vapour leaves the still and passes into a condenser by way of "gooseneck". Gauze or screen is fitted at the mouth of the gooseneck to prevent plant material being blown into the condenser.

The simple form of condenser is shows vapours pass through a coiled tube contained in a water bath and cooled condensate is obtained at the bottom of the condenser tube. It is important that condensation is complete or oil may be lost by evaporation.

A more efficient type of condenser is the multi-tubular type shows a series of parallel tubes are mounted inside a cylindrical jacket though which cooling water is passed. This design provides a large surface cooling in relation to its volume.

The mixture of water and essential oil leaves the condenser and flows into a separator, called a florentine flask, in which they separate into two layers. The essential oil will generally be lighter than water, the oil floats to the surface and the distillate water drains away.

It is important that the oil separators should have a large volume to minimize turbulence because significant amounts of oil can be lost with the distillate water if the oil is not allowed to separate completely. The temperature of the distillate has an important bearing on the efficiency of separation of essential oil and water. The optimum temperature for obtaining the best separation should be found. When separation of oil is difficult, the distillate water is run back into the still (cohobation) and redistilled.

Condensers and separators should be constructed of inert materials which do not react with essential oils or water. Mild steel rusts and is not suitable. Copper has been used for many years (tinned, since copper reacts with the oil). The optimum material for stills, condensers and separators is stainless steel which is resistant, durable, but relatively expensive.

Concrete : An extract of a fresh plant part obtained by extraction using a hydrocarbon solvent. It is rich in perfumery principles soluble in a hydrocarbon solvent and is devoid of water soluble components. It is usually a dark coloured semi –solid waxy material.

Absolute : An alcoholic extract of the concrete containing only alcohol soluble materials.

Pommade : The product obtained by extraction of fresh flowering using highly purified fats. Extraction of pomade with alcohol affords the absolute. The technique is known as enfluerage.

Oleoresin of: Extract of dried spice/ herb using a solvent for extraction. It is used spice/herb in food and pharmaceutical industries as a replacement for a ground spice.

USES OF AROMATIC PLANTS

The term "aromatics", which includes all aromatic botanical substances used in foods, although spices like cinnamon, cloves, mace are in fact a sub-classification, along with aromatic sees (coriander, celery, cumin) and herbs (oregano, bay, thyme). Aromatic plants are used in the following ways :

Seasonings

Include blends of spices, with onion, garlic, MSG, salt, sugar, e.g. chili, curry powder. These are used in 'sprinkles' to season cooked food dishes.

Whole-Spices

Few whole spices are used in food industry, except for appearance in e.g. pickling spices (celery, mustard seeds), bakery products (sesame, poppy sees), and meat industry (black pepper).

Ground-Spices

These have been widely used but are being superseded by more consistent flavour sources. They are used widely in bakery, meat and canned foods. They are available in sizes from cracked seeds to finely milled (10-50 microns) powders. Oily spices such as nutmeg are difficult to grind. Low temperature grinding is commonly used, with dry ice or liquid nitrogen to remove heat and 'set' the spice, lowering the vapour pressure of the essential oil and retaining the volatiles.

Spice-Oils

These are the essential oils (2-15%) and the non-volatile oleoresin. The latter contains pungent or heat principles of the spice (paprika, turmeric, saffron) extracted with low-boiling solvents (e.g. methanol, isopropanol).

Synthetic-Spice-Oils

Many "extenders" to spice oil are known e.g. oil of cruise, cinnamic aldehyde, which are much cheaper than the natural volatiles, and used for this reason, although this is adulteration.

Soluble-Spices

These are derivatives of spices mixed with soluble carriers (sucrose, salt, MSG), using the volatile oil from distillation or oleoresins from solvent extraction. The blend will reflect a truer flavour of the original spice.

Encapsulated-Spices

The modern food manufacturer often uses this form of the spice as it gives a controlled release of the flavour into the food, depending on the solubility of the encapsulating compound (e.g. gelatin).

The food manufacturer nowadays evaluates spices for flavour stability, minimum variability, release, distribution, appearance, absence of contamination, low tannins content, no handling problems identifiable and quality standards.

— xxx —

Chapter 26

Post-Harvest Aspects

Preparation of material for distillation varies with the material to be distilled. Some material must be distilled immediately after harvesting, whereas others are best stored indefinitely before distillation. In general, flowers should be distilled immediately, whereas herbaceous material benefits from wilting for one or two days before distillation. Woody materials need to be ground and/or soaked before distillation.

The preparation of the raw material, the packing of the still and the rate/type of distillation can be determined for a particular essential oil crop from the literature or from experimental trials.

Leaf Materials

Care must be taken in preparing the leafy plant material for drying or distillation. Freshly harvested plants are generally allowed to wilt and lose moisture in the field before transporting, although some volatiles are lost. Between 24-72 hours wilting is issued, but the plant should be kept dry and cool to prevent fermentation or mould growth. A concrete floor under shade can be used. Harvesting is labour intensive and is carried out early morning or evening. The best oil is in the top leaves.

Flowers

Generally, flowers are dried immediately after picking, not in the sun but in shade, turning over to prevent browning. Mechanical driers are used for volume operation. Preservation is achieved with a low moisture content in the flowers.

Storage and Packing

Most essential oils can be stored for long periods under suitable conditions. They should be dry, not in contact with the air or direct sunlight and kept cool. It is important that essential oils do not come into contact with materials with which they might react, e.g. rubber or plastic stoppers.

Glass containers are often used for smaller amounts of oil but larger quantities are stored in metal drums. Mold steel drums lined with epoxy resin are very popular for essential oils. If second-hand drums are to be used, it is important that they are thoroughly cleaned and dried before being filled with essential oil. Plastic containers, e.g. polythene, should not be used because the oil may be absorbed by the plastic and contamination may occur. To ensure that the oil is not wet it should be left to stand for some time before being filtered into its containers. Oils generally show no cloudiness when thoroughly dry.

Freshly distilled oils often possess some "still odours" which are unpleasant. These generally disappear after several weeks storage. Some oils gradually improve in storage and acquire a fuller, more rounded aroma, e.g. vetiver and patchouli.

The common types of packaging materials currently available are given below (Krainsintu, 1997).

1. *Glass* : It can be found as several variants, such as, treated soda glass, soda glass and non-parenteral.
2. *Metals* : A variety of metals including tin plate (tin-coated mild steel) tin-free steel, aluminium, aluminium alloys are widely used in packaging, being found as rigid containers, collapsible containers, aluminium foils, metalized coatings, etc.
3. *Plastics* : There are five economical materials for rigid type of containers i.e., those based on polyethylene (PE), polypropylene (PP), polystyrene (PS), PVC and polyester.
4. *Elastomeric materials* : Elastomers can be found as a wide range of basic materials (i.e. natural rubber, synthetic polyisoprene, neoprene, nitryl, butyl, including bromo and chloro-butyl, ethylene propylene diene modified (EPDM), acid silicone elastomers).

Post-Harvest Processing of Some Medicinal & Aromatic Plants

S. No.	Medicinal Plant	Useful part	Post Harvest Processing
1.	Mulathi	Root	After digging cut the root upto 5-20 cm and dry it by spreading under shade till its moisture content in 10 years (dry under 30-40°C).
2.	Kalihari	Seed	When its fruit become slight green in colour and skin becomes shrinked then pluck it

Contd...

Teniposide	Bladder neoplasms	*Podophyllum peltatum*
Tetrahydroc-annibinol	Antiemetic	*Cannabis sativa*
Theophyllin	Diuretic, antiasthmatic	*Camellia sinensis*
Toxiferine	Relaxant in surgery	*Strychnos guianensis*
Vinblastine	Hodgkin's disease	*Catharanthus roseus*
Vincristine	Paediatric leukaemia	*C. roseus*
Xanthotoxin	Vitiligo	*Ammi majus*

Method for Preparation of Some Value Added Products

1. Amritdhara

Requirement

Peppermint	-	10gm
Thymol	-	10gm
Kapoor (Bhimseri)	-	10gm

Method

(i) All the ingredients are put into a glass bottle and mixed well. After some time all ingredients are convert into liquid form.

2. Herbal Tea

Requirement

Arjun	-	100gm
Kalimirch	-	5gm
Nagarmotha	-	10gm
Lalchandan	-	10gm
Mulethi	-	10gm
Tezpatta	-	10gm
Dalchini	-	10gm
Tulsipatr	-	10gm

Method

(i) Clean all the ingredients and grind them to a fine powder and store into a glass bottle for use.

3. Bhringraj Hair Oil

Requirement

Refined coconut oil	-	500gm
Bhringraj (fresh) leaves	-	250gm

Method

(i) Heat Coconut oil in a iron vessel on heater.

(ii) Add clean bhringraj fresh leaves when the oil become half and allow leaves to fry completely.

(iii) Now the oil vessel is removed from heater and allows it to cool down and there after it will be filter by a muslin cloth.

4. Herbal Tooth Powder

Requirement

Turmeric	-	100gm
Salt	-	100gm
Triphla	-	100gm
Arjun powder	-	100gm
Neem leaves	-	100gm
Clove	-	20gm

Method

(i) Take all ingredient separately and clean them.

(ii) Grind all of them separately to fine powder except alum which will be heated till it will become fine powder.

(iii) Mix all the powder and store in a glass bottle for daily use.

5. Chyavanprash

Requirement

Giloy	-	500gm
Jeevanti	-	10gm
Cacarmool	-	10gm
Satavar	-	20gm
Ashwagandha	-	20gm
Gokhur	-	10gm
Badi harr	-	10gm
Munakka	-	10gm
Bhui amla	-	10gm
Safed musli	-	20gm
Kakamachi	-	10gm
Astvarg	-	50gm
Nagarmotha	-	10gm
Dashmool	-	50gm
Cardmom	-	10gm
Safed chandan	-	10gm
Pipalee	-	10gm
Amla	-	1kg
Water	-	3ltr
Sugar	-	1kg
Honey	-	200gm
Ghee	-	250gm

Method

(i) Mixed all the material except cardamom, pipalee and boil with 3 liter of water kept the anvala in cloth and tie it, put this with all the above material.

(ii) When anvala material become left than remove it and allows only water to boil till it reduces to about 300ml their after filter the same.

(iii) Now make a paste, boil anvala in a mixer after that put anvala paste in a hot pure ghee in a stainless vessel and than add filtered water ghee start leaving the vessel.

(iv) Confirmly stair with large wooden spoon and add sugar powder or misry powder, elaichi and pippali stair the whole for about 10 minute.

(v) Now allow it to cool and then add honey to this finally pour it in a glass or plastic bottle.

contd....

Glossary of Medical Terms

Analgesic	Relieves pain
Acrid	Helps rhermogenic and metabolic processes
Alexipharmic	Antidote to poison
Alexiteric	Protects from infectious diseases
Alkaloid and Trigonelline	Improve glucose oxidation
Alkoids	Reduces absorption of additional glucose
Anodyne	Reduces pain
Anthelmintic	Destroys intestinal worms, Extricates worms from intestine
Anthelmintic	Kills the worms
Anti anaemic	Increases the hemoglobin level
Anti bacterial	Antiseptic
Anti cholesterol	Controls cholesterol level
Anti diabetic	Purifies blood and improves peripheral glucose uptake

Purgative	Improves bowel movement generally affected due to reduction of body movement
Refrigerant	Increases cooling effect
Regenerative	Regenerates or repairs activity of pancreas, improves functioning of liver
Rejuvenating	Builds immunity
Roborant	Strengthens the body
Saponins and Tritertines	Reduces cholesterol level
Sedative	Calms the nerves
Sialogogue	Increases the flow of saliva
Soothing	Relaxing
Spleenomegaly	Cures inflammation of spleen
Stimulant	Acts as nerve tonic, Stimulates and strengthens the immune system
Stomachic	Increases appetite
Supportive	Nourished the body
Thermogenic	Generates heat and reduces swelling and joint pain, clean the ear, provides strength to the body
Tranquilising	Calms, sooths and pacifies effect without depression
Trichogenous	Helps in blackening and strengthening of hair
Vermifuge	Kills the worms
Vulnerary	Promotes healing of wounds

— xxx —

Bibliography

Azhar Ali Farooqi and B.S. Sreeramu., (2001). Cultivation of Medicinal and Aromatic Crop. Universities Press (India) Limited, Hyderabad.

C.M., Franz, (1993). Personal Communication. Address: Institute for Botany and Food Science, University of Veterinary Sciences, Links Bahngasse 11, A103, Wien, Austria.

K. P. S. Chandel, G. Shukla & N. Sharma., (1996), Biodiversity In Medicinal And Aromatic Plants In India: *Conservation and Utilization*. National Bureau of Plant Genetic Resources, New Delhi, India.

K. Kraisintu, (1997). Industrial Exploitation of Indigenous Medicinal and Aromatic Plants: Formulation and Industrial Utilization. UNDP.

N.R. Farnsworth, (1984). The role of medicinal plants in drug development. Natural products and Drug Development Krogsgaard-Larsen P, Christensen S B,Kolod H,eds. Munksgaard, Copenhagen,17-28.

P.P. Joy, J. Thomas, S. Mathew and B.P. Skaria., (2001). Tropical Horticulture (Medicinal Plants) Naya Prakash Calcutta, India.

P.P. Joy, J. Thomas, S. Mathew, G.Jose and J. Joseph., (2001). Tropical Horticulture (Aromatic Plants). Naya Prakash Calcutta, India.

Pulok K. Mukharjee., (2002). Quality Control Herbal Drugs An Approach to Evaluation of Botanicals. Business Horizons, New Delhi, India.

Ravindra Sharma., (2004). Agro-Techniques of Medicinal Plants. Daya Publishing House.

R.O.B. Wijesekera, (1993). Technological Variations in the Essential Oils Industry, Proc.2nd UNIDO Workshop on the Essential Oils Industry, Manila, WESOIN 2.IO/R259, UNIDO, Vienna; pp.16-45.

Tej Verma., (2003). Medicinal plants for Health Care, AICRP on Home Science (Extension component) Indian Council of Agricultural Research, New Delhi, India.

Tej Verma., (2003). Medicinal plants for Digestive Disorders, AICRP on Home Science (Extension component) Indian Council of Agricultural Research, New Delhi, India.

Tej Verma., (2003). Medicinal plants for Respiratory Disorders, AICRP on Home Science (Extension component) Indian Council of Agricultural Research, New Delhi, India.

Tej Verma., (2003). Medicinal plants for Aliments and Diseases, AICRP on Home Science (Extension component) Indian Council of Agricultural Research, New Delhi, India.

— xxx —

Lahsun

Neem

Jira

Haldi

Laung

Ghi Kanvar

PLATE - 1

Buch

Madar

Motha

Guggal

Datura

Amla

PLATE - 2

Karanj

Pipali

Aphim

Arandi

Bariarai

Makoi

PLATE - 3

Senna

Brahmi

Bargad

Satavar

Sounf

Ghirayita

PLATE - 6

Giloy

Hari Mirch

Isabgol

Adrak

Mulathee

Kali Mirch

PLATE - 7

Bhang

Babul

Bahera

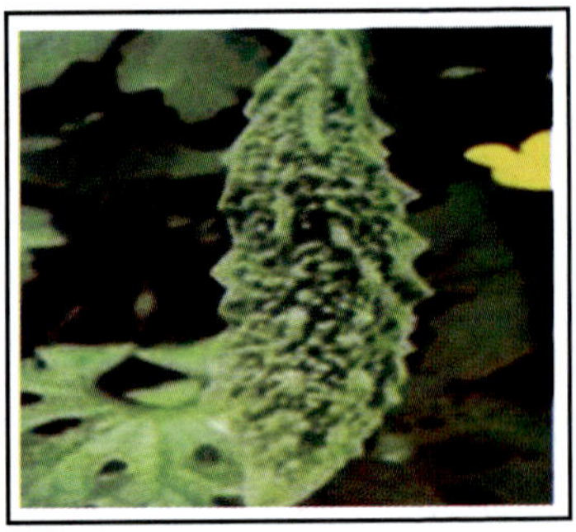

Karela

Eucalyptus

Kateli

PLATE - 8

Ghatwan

Gudmar

Aam

Bijasal

Jamun

Harra

PLATE - 9

Sadabahar

Pudina

Bael

Dhania

Imli

Ashok

PLATE - 10

Dhak

Henna

Babul Tulsi

Kesar

Jamal Gota

Adusa

PLATE - 11

Kagzi Nimbu

Isharmul

Ashwagandha

Sarpagandha

Kalmegh

Kattha

PLATE - 12

Punarnava

Bhringraja

Kapuri

Bhui Amla

Chitrak

Harra

PLATE - 13

Rai

Arhar

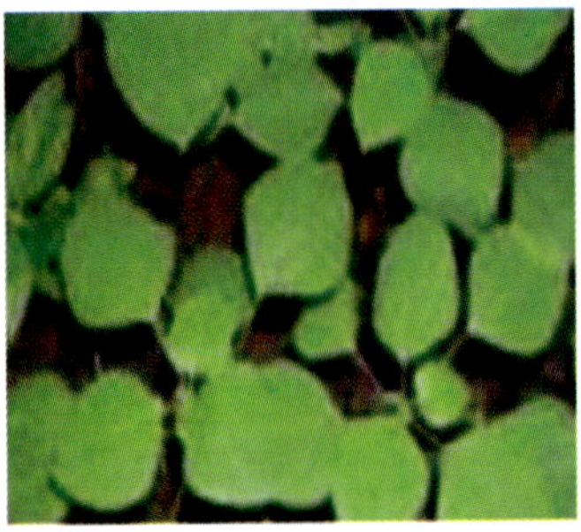

Kupi Khokli

Gudmar

Bhringraja

Aravi

PLATE - 14

Ritha

Kachnar

Manduka Parni

Tarwar

Meethi Neem

Vidanga

PLATE - 15

Gajar

Jamun

Pipal

Sowa

PLATE - 16

❑❑❑

ABOUT THE AUTHORS

Dr. (Mrs.) Anubha Upadhyaya, Ph.D (Food and Nutrition) is Associate (Food Science). Last 7 years she is working in Department of Crop and Herbal Physiology, Collage of Agriculture, Jawaharlal Nehru Krishi Vishwa Vidayalaya, Jabalpur, India. She started her career in JNKVV as Assistant Professor, has been working in the field of research extension and teaching. She is a specialist of bioanalytical separation techniques and an expertise in developing and standardizing procedures through modern chromatographic instrumentation like GC, GLC, HPLC, HPTLC, etc. She has guided many Ph.D and M.Sc students and has more than 30 papers published in international and national journals. She has been actively involved in giving training to farmers and technical person in the field of Post Harvest Technology and Value added products of Medicinal and Aromatic plants. She is Co-PI in the Project Niche Area of Excellence in Conservation, Cultivation, Processing and Quality evaluation Medicinal and Aromatic plants and Quality analysis of Medicinal and Aromatic plants.

Dr. (Mrs.) Preeti Sagar Nayak, Ph.D (Chemistry) is employed as a Research Associate in the Department of Crop and Herbal Physiology, College of Agriculture, Jawaharlal Nehru Krishi Vishwa Vidayalaya, Jabalpur, India, since early 2006. For her doctoral degree at Rani Durgavati University Jabalpur, her research focused on sorption and techniques for understanding of chemistry with an emphasis on developing more accurate separation and remedial tools. At present her work is related to qualitative and quantitative isolation / extraction of bioactive component of Medicinal and Aromatic plants using chromatography. In addition to her research and publications in physical and bio-analytical chemistry, she has distinguished herself as an outstanding research worker. Dr. Nayak is a vibrant and enthusiastic young scientist possessing highly developed intellect. She has widely travelled in India and has attended many symposia and conferences for the purpose of presenting her research work. She has in her credit more than 12 research papers of international and national repute.

Dr. V.K.Agrawal, Ph.D (Plant Physiologist) is a Professor and Head, Department of Crop and Herbal Physiology, College of Agriculture,

Jawaharlal Nehru Krishi Vishwa Vidayalaya, Jabalpur, Madhya Pradesh, India. Basically he is a plant physiologist having research, teaching and extension experience of more than 35 years. He is leader of Medicinal and Aromatic plants R & D unit at JNKVV, Jabalpur that has name and fame in collecting, conserving and domesticating valuable herbal biodiversity not only in the state but also in the country. He is principal investigator of the project financed by ICAR entitled, Project Niche Area of Excellence in Conservation, Cultivation, Processing and Quality evaluation Medicinal and Aromatic plants. About fifty research papers and many books and popular articles have been contributed by him in the field of Medicinal and Aromatic plants. He has been instrumental in organizing seminars, trainings, workshop and symposia on Medicinal and Aromatic plants. He is playing key role in introducing many of these plants into agriculture studies. Cultural and cultivation practices are being undertaken for those plants which are found suitable and remunerative for commercial cultivation.

Shri. A.B.Tiwari, (Taxonomist) is Scientist in Department of Crop and Herbal Physiology, College of Agriculture, Jawaharlal Nehru Krishi Vishwa Vidayalaya, Jabalpur, India. He has vast knowledge of Medicinal and Aromatic plants related to their taxonomy, ethnobotany, healing and pharmaceutical properties. He is Co-PI of many central and state funded research projects. He is author of many extension bulletins and research papers. Cultural and cultivation practices are being undertaken for those plants which are found suitable and remunerative for commercial cultivation.